GRADE 5

SCHOLASTIC

New York • Toronto • London • Auckland • Sydney
Mexico City • New Delhi • Hong Kong • Buenos Aires

Teaching Resources

State Standards Correlations

To find out how this book helps you meet your state's standards, log on to **www.scholastic.com/ssw**

Cover design by Ka-Yeon Kim-Li
Interior design by Ellen Matlach Hassell
for Boultinghouse & Boultinghouse, Inc.

ISBN-13 978-0-545-20067-7
ISBN-10 0-545-20067-9

4 5 6 7 8 9 10 40 17 16 15 14 13 12 11

Contents

"Nothing succeeds like success."
—Alexandre Dumas the Elder, 1854

And no other math resource helps kids succeed like Scholastic Success With Math! For classroom or at-home use, this exciting series for kids in grades 1 through 5 provides invaluable reinforcement and practice for math skills such as:

- ❑ number sense and concepts
- ❑ reasoning and logic
- ❑ basic operations and computations
- ❑ story problems and equations
- ❑ time, money, and measurement
- ❑ fractions, decimals, and percentages
- ❑ geometry and basic shapes
- ❑ graphs, charts, tables . . . and more!

Each 64-page book contains loads of challenging puzzles, inviting games, and clever practice pages to keep kids delighted and excited as they strengthen their basic math skills.

What makes *Scholastic Success With Math* so solid?

Each practice page in the series reinforces a specific, age-appropriate skill as outlined in one or more of the following standardized tests:

- ***Iowa Tests of Basic Skills***
- ***California Tests of Basic Skills***
- ***California Achievement Test***
- ***Metropolitan Achievement Test***
- ***Stanford Achievement Test***

These are the skills that help kids succeed in daily math work and on standardized achievement tests. And the handy Instant Skills Index at the back of every book helps you succeed in zeroing in on the skills your kids need most!

Take the lead and help kids succeed with *Scholastic Success With Math*.
Parents and teachers agree: No one helps kids succeed like Scholastic!

What's in a Word?

Name ______________________________ Date ______________

✏ A prefix is a word part added at the beginning of a word. A prefix changes the meaning of a word. The prefixes in this activity help form words that represent numbers. Each statement contains a word with a number prefix. The list below contains numbers written out as words. Fill each blank with the correct word from the list below.

1. An animal with ____________ horn on its head is called a **uni**corn.
2. A **dec**ade lasts ____________ years.
3. An **oct**opus has ____________ tentacles.
4. A **tri**athlete participates in ____________ Olympic events.
5. A **bi**cycle has ____________ wheels.
6. A **cent**ury marks a ____________ years.
7. A **non**agon is a shape with ____________ sides.
8. A **kilo**meter is equal to a ____________ meters.

nine **ten**
hundred **three**
two **eight**
one **thousand**

Research other number prefixes. Try finding some that represent larger numbers. Share them with the class.

Cat Stats

Name ______________________________ Date ____________

✏ Every year, cats from all over the world come to Kentucky to participate in the Cat Club's Annual Crazy Costume Contest. Some cats come dressed as their favorite people. Others dress up like other animals. It was a tough decision, but the judges have found their winner. Do you know which cat won?

DIRECTIONS:

Use the scores next to each contestant's name to find their average score. Write the averages in the spaces provided. The contestant with the highest score is the winner.

CATS	SCORE	AVERAGE
Sabrina Siamese	16, 11, 15, 18	______
Freddy Feline	10, 12, 14, 16	______
Karl Kat	14, 15, 17, 18	______
Kelly Kitten	18, 14, 12, 12	______

Which cat is the winner? ______________

Who came in second? ______________

Third? ______________

Fourth? ______________

Try coming up with your own costume or talent contest. Choose four or five judges who will score each contestant on a scale of 10-20. Find the average of each contestant's scores to come up with the winner.

Pig Patterns

Name ______________________________ Date ______________

Riddle: What would you get if a pig learned karate?

Find the answer by completing the next step in the pattern. Then use the Decoder to solve the riddle by filling in the blanks at the bottom of the page.

1. 2, 4, 6, 8, ___
2. 1, 3, 5, 7, ___
3. 3, 7, 11, 15, ___
4. 5, 10, 15, 20, ___
5. 10, 20, 40, 80, ___
6. 1, 5, 3, 7, 5, 9, ___
7. 15, 25, 20, 30, 25, ___
8. 0, 1, 3, 6, 10, ___
9. 9, 18, 36, 72, ___
10. 100, 200, 100, 300, 100, ___

Decoder

12	D
160	P
96	T
10	R
20	F
400	H
40	G
25	O
35	S
19	C
500	B
11	K
7	E
16	W
9	O
21	A
15	P
30	I
144	K

SOM ___ ___ ___ ___ ___ ___ ___ ___ ___ ___

6   8 4 1 9   3 10 2 5 7

A Stinky Riddle

Name ________________________ Date ____________

Riddle: How do skunks measure length?

Answer each problem. Then use the Decoder to solve the riddle by filling in the spaces at the bottom of the page.

1. In the number 52,370, the digit 2 is in which place? _______
2. In the number 619,246, which digit is in the hundred thousands place? _______
3. In the number 2,027,635, the digit 3 is in which place? _______
4. In the number 37,196,511, which digit is in the millions place? _______
5. In the number 402,819,335, which digit is in the ten millions place? _______
6. In the number 9,817,248,100, which place is the digit 9 in? _______
7. In the number 6,543,210,789, which place is the digit 5 in? _______
8. Which number is greater: 727,912 or 699,534? _______
9. Which number is smaller: 4,847,266 or 5,000,122? _______
10. Which number is greater: 7,446,726,012 or 7,446,732,011? _______

Decoder

7,446,726,012 .... **K**
ones.................. **P**
1........................ **W**
4,847,266 .... **T**
7........................ **N**
thousands .... **I**
699,534 .... **A**
hundreds............ **O**
7,446,732,011 ..... **T**
billions.............. **R**
tens .... **S**
ten millions .... **B**
6........................ **E**
5,000,122 .... **D**
ten thousands..... **V**
0........................ **E**
hundred millions ..**M**
9........................ **F**
5........................ **H**
727,912 .... **E**

IN "SC ___ ___ ___" ___ ___ ___ ___ ___ ___ ___

8 4 9 1 7 5 10 2 6 3

Talented Tongue

Name ______________________________ Date ______________

Figure It Out!

1. Using RIBBIT and CROAK, a frog can make these 2-word phrases: RIBBIT-CROAK and CROAK-RIBBIT. What 2-word phrases can a dog make of BARK and RUFF? (Use each word only once in each phrase.) ______________

2. How many different 2-word phrases can a dog make out of the words BARK and GRR? Write each arrangement. ______________

____

3. How many different 2-word phrases can a dog make out of the words BARK, GRR, and RUFF? Write each arrangement. ______________

____

4. How many different 2-word phrases can a cat make out of the words MEOW, PURR, and SSS? Write each arrangement. ______________

____

5. How many different 3-word phrases can a cat make out of the words MEOW, PURR, and SSS? Write each arrangement. ______________

____

SUPER CHALLENGE: How many different 3-word phrases could a cat make out of the words MEOW, PURR, and SSS if each phrase must start with the word PURR?

Cow Rounding

Name ______________________ Date ____________

Riddle: What do cows give after an earthquake?

Decoder

700	F
11,000	K
800	S
2,780	O
3,600	U
1,000	M
9,900	Y
24,400	I
73,000	S
5,000	L
24,000	P
6,000	Q
2,770	E
7,500	T
9,940	A
3,700	K
10,000	R
8,000	H
2,000	N

Round each number. Then use the Decoder to solve the riddle by filling in the spaces at the bottom of the page.

1. Round 789 to the nearest hundred ________
2. Round 5,112 to the nearest thousand ________
3. Round 3,660 to the nearest hundred ________
4. Round 1,499 to the nearest thousand ________
5. Round 2,771 to the nearest ten ________
6. Round 7,529 to the nearest thousand ________
7. Round 24,397 to the nearest hundred ________
8. Round 10,708 to the nearest thousand ________
9. Round 9,937 to the nearest ten ________
10. Round 73,489 to the nearest thousand ________

__ __ __ __   __ __ __ __ __ __

4 7 2 8   10 6 9 3 5 1

Apple Add-Up

Name ________________________________ Date ______________

Object: To cover more apples than the other player.

Number of Players: 2

You Need:
- small counters
- pencil
- paper clip

To Play:

- Each player gets a copy of the apple tree game board. Decide who will go first.
- Take turns spinning. (Look at the picture to see how to use the spinner.) After spinning, cover **two** apples on your tree with the counters. The two numbers on the apples must add up to a number that matches the spinner. **Example:** Player 1 spins "Equal to 15." Player 1 can cover 7 and 8, 5 and 10, 2 and 13, or any other combination of two apples that totals 15. Player 2 spins "More than 15." Player 2 can cover any combination of two apples that totals more than 15.
- Once your counters are on the board, you can't move them!
- If you can't cover two apples to match the spinner, you're out. The other player wins.

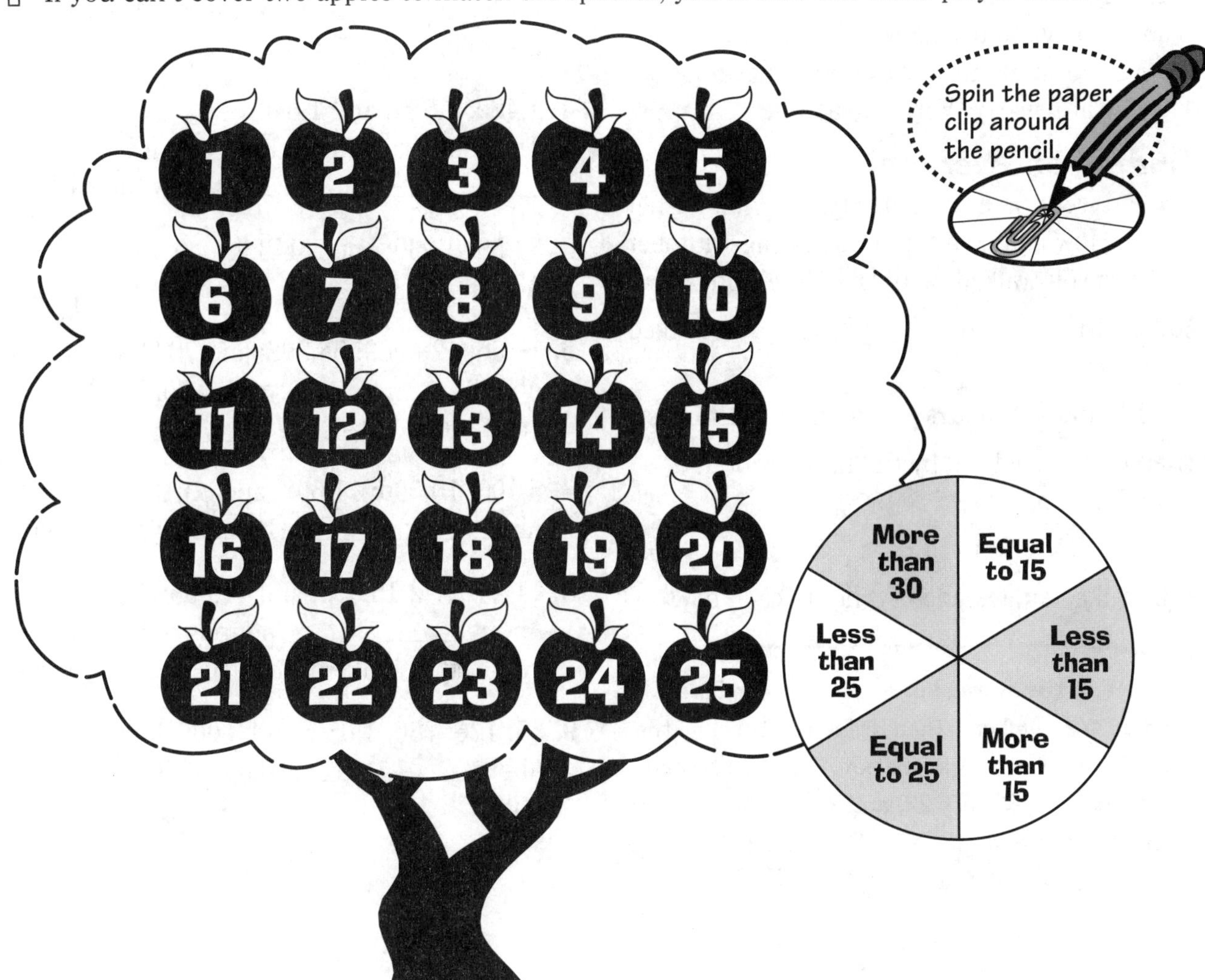

A "World" of Averages

Name ______________________________ Date ______________

For many people, Florida's Walt Disney World is a magical place. You might say there's nothing average about it. But if you look closely, you can find lots of averages there!

What's an average? It's a number that describes a group of numbers. It isn't the biggest number in the group, or the smallest. It's somewhere in between. For example, the average number of people that visit Walt Disney World each day is about 77,000.

That doesn't mean that exactly 77,000 people visit the park every day. On a sunny day or a holiday, more than 77,000 people might visit the park. On a rainy day, fewer than 77,000 people might visit. But 77,000—the average—is about how many people visit on most days.

Want to find out more about an average day at Walt Disney World? Read on!

Finding an Average

Say you went on a three-day trip to Walt Disney World. How could you find the average number of hours you walked each day? Here's one way:

Add up the actual number of hours you walked each day:

10 hours + 8 hours + 6 hours = 24 hours

Then divide the total by the number of days you added up.

24 hours ÷ 3 days = 8 hours

You walked an average of 8 hours each day.

To find the average of any set of numbers, add all the numbers. Then divide the total by the number of numbers in the set. Example: to find the average of 40, 30, 22, and 20, first add. Then divide the total, 112, by 4. The average is 28.

What to Do:

By finding the average of each set of numbers below, learn more about what happens on an "average" day at Walt Disney World.

1. 25 and 175
About __________ pairs of sunglasses are turned in to the Lost and Found in the Magic Kingdom every day.

2. 5,000 and 7,000
You can choose from about __________ different food items.

3. 881; 924; and 1,234
About __________ Mickey Mouse ears are sold.

4. 1,489; 1,584; and 1,640
The monorail trains travel about __________ miles in and out of the parks.

5. 3,259; 4,039; and 5,443
About __________ T-shirts are bought.

6. 10,660; 28,069; 58,392; and 78,223
About __________ packets of ketchup are handed out.

7. 5,400; 10,000; 11,608; and 33,124
About __________ hamburgers are sold.

8. 117; 3,274; 15,673; and 41,208
About __________ pounds of potatoes are used to make french fries.

9. 35; 126; 780; 1,050; and 3,009
About __________ Band-Aids are given out.

All Mixed Up

Name ______________________ Date ____________

Finding the sums is easy. But when you try to put these numbers correctly in the puzzle, you'll find yourself all mixed up!

Find the sum and write the answer in the puzzle. Each digit can occupy only one place to make the whole puzzle fit together perfectly. The first one has been done for you.

54 + 98 152	69 + 37	31 + 85	292 + 614	589 + 92	261 + 97
423 + 79	180 + 98	349 + 301	2,012 + 2,106	413 + 923	855 + 723
1,617 + 1,281	4,068 + 784	1,602 + 639	5,142 + 2,690	1,069 + 1,103	1,597 + 346
4,115 + 106	1,022 + 1,886	951 + 1,384	12,401 + 6,001	44,595 + 13,816	5,354 + 1,346

Calculate a Happy Chinese New Year

Name ______________________ Date __________

Everyone loves to celebrate New Year's Eve. But once January 1 is past, you don't need to hang up your party hat. You can still celebrate Chinese New Year, which falls in late January or February.

On the Chinese calendar, each year is named after one of 12 animals. For example, 1999 was the Year of the Rabbit. 2005 was the Year of the Rooster. The animals are all on your game board.

What to Do:

- Use pages 14 and 15 for this game.
- Pick any number and write it in the first rooster, by START.
- Follow the animals around the game path. Do what their sign and number tell you. Keep track of your total on scratch paper.
- Every time you get to a rooster, you should get the number you started with as the answer. If you don't, go back and check your work.
- After you finish, play again with a different number. Want a challenge? Try using a 3-digit number.

Which Animal Are You?

Find the year you were born on this chart. Which animal sign were you born under? Which animals are others in your family?

Animal	Year	Animal	Year
Cow	1997	Sheep	2003
Tiger	1998	Monkey	2004
Rabbit	1999	Rooster	2005
Dragon	2000	Dog	2006
Snake	2001	Pig	2007
Horse	2002	Mouse	2008

A Riddle to Grow On

Name ______________________ Date ____________

Riddle: What tables grow on farms?

Do each subtraction problem. Then use the Decoder to solve the riddle by filling in the spaces at the bottom of the page.

1. 714 – 457 = ________
2. 936 – 508 = ________
3. 1,000 – 700 = ________
4. 1,362 – 619 = ________
5. 2,000 – 549 = ________
6. 3,873 – 1,004 = ________
7. 1,446 – 987 = ________
8. 5,011 – 4,963 = ________
9. 8,600 – 3,716 = ________
10. 9,925 – 1,999 = ________

Decoder

4,884	T
64	C
275	D
459	V
286	W
1,451	B
257	L
1,541	K
428	G
81	M
743	E
48	E
792	P
2,869	S
12	Z
300	E
2,942	Y
7,926	A
7,431	Q

" ___ ___ ___ ___ " ___ ___ ___ ___ ___ ___

7 4 2 8 9 10 5 1 3 6

Multiplying & Dividing

Name ______________________ Date ____________

Using the digits in the box, write the answer to each number riddle in the form of an equation. Digits appear only once in an answer.

8 1 4 7 3

1 The product of a 1-digit number and a 2-digit number is 284.

What are the numbers? ______________

2 The product of two 2-digit numbers, plus a number, is 3,355.

What are the numbers? ______________

3 The product of a 3-digit number and a 1-digit number, minus another 1-digit number, is 1,137.

What are the numbers? ______________

4 The product of a 2-digit number and a 3-digit number is between 13,000 and 14,000.

What are the numbers? ______________

5 When a 3-digit number is divided by a 2-digit number, the quotient is between 5 and 6.

What are the numbers? ______________

6 When a 2-digit prime number is divided by another 2-digit prime number, the quotient is nearly 5.

What are the numbers? ______________

Picture-Perfect Star

Name ______________________________ Date ____________

Solve the problems. Then color the design. Here's how: **1.** Choose four colors that you like. **2.** Write the name of one of the colors on each line below. **3.** Color the puzzle. If the answer is between 1 and 200, color the shape ____________. If the answer is between 201 and 500, color the shape ____________. If the answer is between 501 and 700, color the shape ____________. If the answer is between 701 and 900, color the shape ____________.

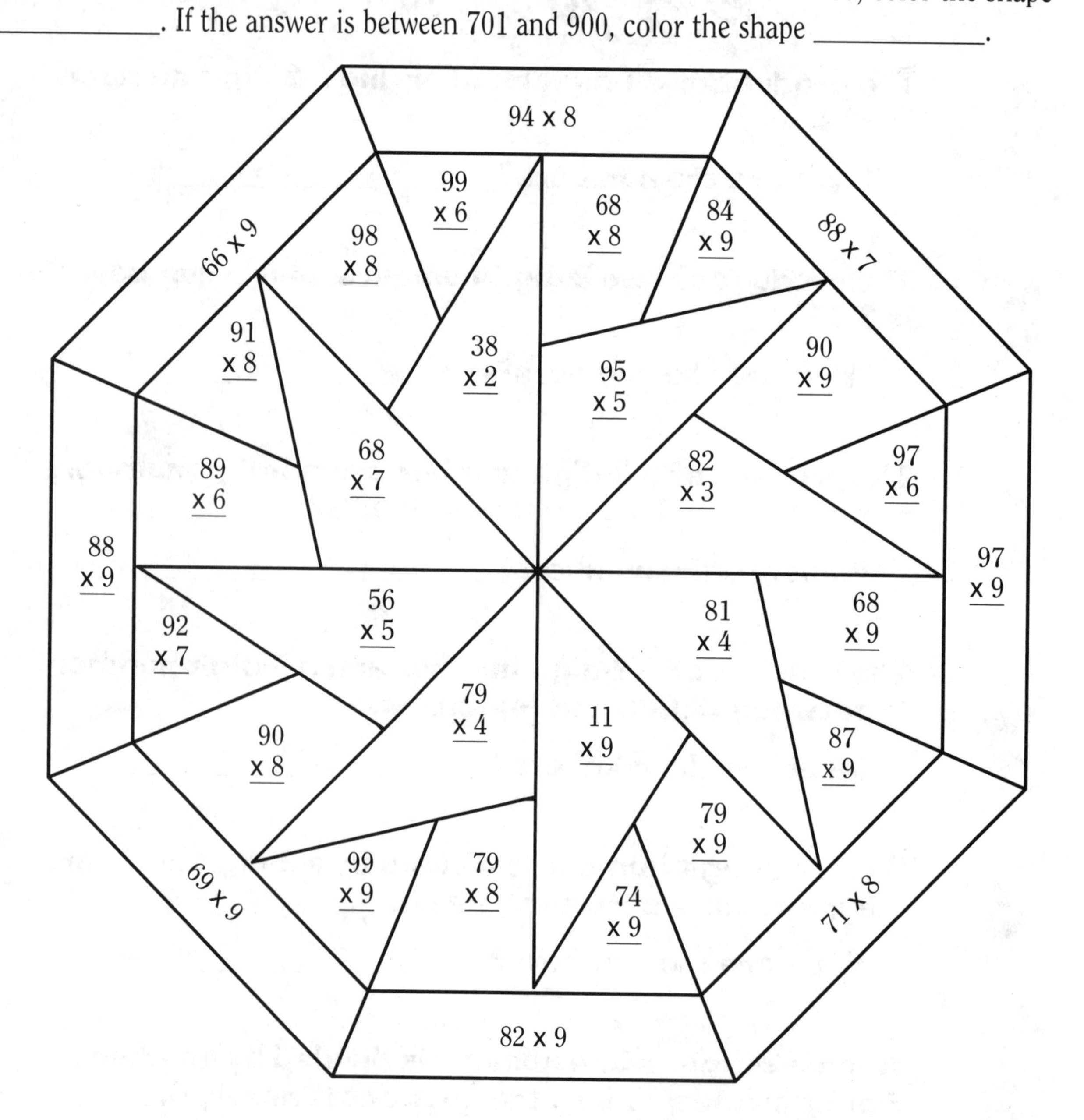

Taking It Further: Fill in the missing numbers.

a. 88 x ______ = 792 b. 56 x ______ = 392 c. 41 x ______ = 246

A "Barber"ous Riddle

Name ______________________ Date ____________

Riddle: How did the detective find the missing barber?

To find the answer to the riddle, solve the multiplication problems. Then, match each product with a letter in the Key below. Write the correct letters on the blanks below.

1. 1 x 2 x 3 = ________
2. 2 x 4 x 1 = ________
3. 5 x 3 x 4 = ________
4. 3 x 7 x 3 = ________
5. 8 x 4 x 5 = ________
6. 6 x 6 x 7 = ________
7. 9 x 2 x 5 = ________
8. 1 x 8 x 7 = ________
9. 7 x 9 x 5 = ________
10. 4 x 6 x 4 = ________

Key

150	V	315	N	225	A
8	E	252	O	63	E
84	K	6	B	90	D
56	W	351	Z	60	T
160	H	96	T	57	X

Riddle Answer: **HE "COM** ___ ___ ___" ___ ___ ___ ___ ___ ___ ___.

1 4 7 10 5 2 3 6 8 9

Fact Search

Name ______________________________ Date ______________

The puzzle below has many hidden multiplication number sentences. You'll find number sentences going across, up and down, and at an angle. Most number sentences overlap. Loop each multiplication number sentence you find. Happy searching!

6	5	0	70	24	1	8	8	64
8	0	7	1	6	12	1	7	7
48	10	0	3	4	7	8	56	6
90	10	9	3	7	21	42	21	7
9	100	5	16	28	5	7	35	42
2	4	8	50	8	3	24	5	3
20	2	40	10	4	6	6	36	35
4	6	80	2	32	18	9	18	4
7	7	49	20	5	4	54	9	6
11	42	28	2	8	10	36	2	24

Caught in the Web

Name ______________________________ Date ______________

Riddle: Why did the spider join the baseball team?

To find the answer to the riddle, solve the multiplication problems. Then, match each product with a letter in the Key below. Write the correct letters on the blanks below.

1. **1,000 x 11** = __________
2. **2,000 x 12** = __________
3. **3,000 x 10** = __________
4. **4,000 x 14** = __________
5. **5,000 x 20** = __________
6. **6,000 x 24** = __________
7. **7,000 x 30** = __________
8. **8,000 x 32** = __________
9. **9,000 x 40** = __________
10. **7,500 x 50** = __________

Key

56,000	H	65,000	M	30,000	C
11,000	I	144,000	T	375,000	C
265,000	B	25,000	N	10,000	Y
360,000	F	256,000	L	100,000	A
210,000	E	90,000	Q	24,000	S

Riddle Answer: TO ___ ___ ___ ___ ___ "___ ___ ___ ___ ___"

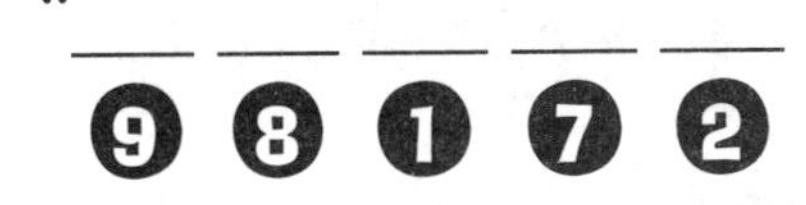

Face Facts

Name ______________________________ Date ______________

✏ The most common way people recognize each other is by the way they look. Each person has distinct eyes, ears, and other features that set them apart from everyone else. Try answering the wacky riddle below to name another feature people have that sets them apart. Factors will help you find the answer.

DIRECTIONS:

Each number is followed by two possible factors. Circle the letter after the number that is a factor. Write the letters in order from the first problem to the last to solve the riddle.

Problem	Number	Factor	Factor
1.	22	2 **I**	6 **O**
2.	70	5 **T**	15 **A**
3.	48	5 **O**	8 **S**
4.	80	16 **T**	11 **B**
5.	120	3 **U**	13 **E**
6.	644	9 **D**	7 **L**
7.	182	13 **I**	4 **L**
8.	156	16 **F**	4 **P**
9.	198	10 **R**	9 **S**

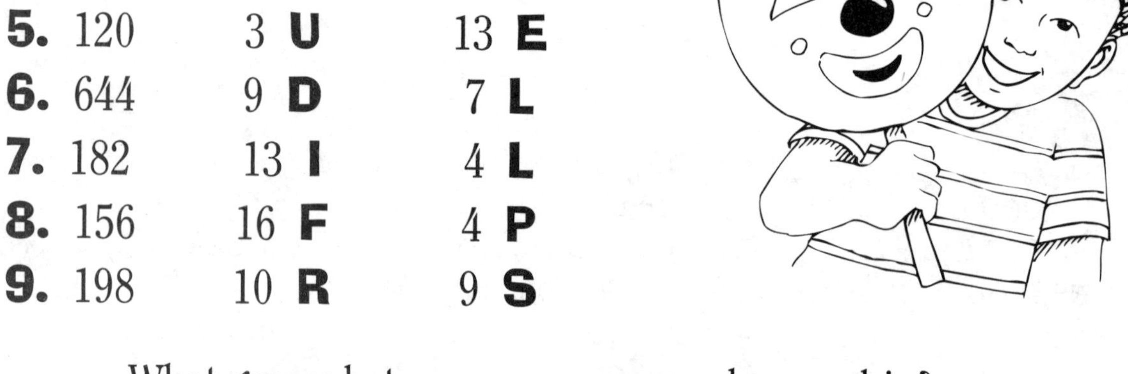

What grows between your nose and your chin?

__ __'__ "__ __ __ __ __ __"!

Make a list of numbers. Ask someone in your class to find at least two factors for each number.

Bug Out!

Name ______________________________ Date ______________

Riddle: What has 18 legs and catches flies?

Find each quotient. Then use the Decoder to solve the riddle by filling in the spaces at the bottom of the page.

1. 74 ÷ 5 = ____________
2. 26 ÷ 9 = ____________
3. 41 ÷ 4 = ____________
4. 55 ÷ 10 = ____________
5. 37 ÷ 14 = ____________
6. 66 ÷ 22 = ____________
7. 84 ÷ 17 = ____________
8. 100 ÷ 11 = ____________
9. 128 ÷ 32 = ____________
10. 200 ÷ 25 = ____________

Decoder

14 remainder 4 . **M**
4 remainder 16.. **L**
5 .... **P**
6 .... **O**
9 remainder 1... **E**
10 remainder 1. **A**
5 remainder 5.... **T**
14 remainder 3. **K**
9 remainder 3... **S**
4 .... **L**
7 .... **C**
2 remainder 8... **S**
4 remainder 15. **N**
8 .... **E**
10 remainder 4. **D**
12 remainder 2. **U**
2 remainder 9... **A**
5 remainder 6... **R**
3 .... **B**

A BA ___ ___ ___ ___ ___ ___   ___ ___ ___ ___

2 10 6 3 7 9   4 8 5 1

Running Riddle

Name ______________________ Date __________

Riddle: What has 3 feet but can't run?

Decoder

20 remainder 10.. **R**
8........................ **B**
30 remainder 40 . **A**
7 remainder 9..... **F**
11...................... **K**
6 remainder 56... **E**
40 remainder 30 **O**
4......................... **T**
12....................... **M**
33 remainder 12.. **I**
32 remainder 12. **L**
9 remainder 7..... **Y**
8 remainder 50.. **N**
30 remainder 23 **D**
9 remainder 50.. **C**
6 remainder 58.. **S**
5 remainder 2.... **Q**
6........................ **W**
5........................ **A**

Find each quotient. Then use the Decoder to solve the riddle by filling in the spaces at the bottom of the page.

1. $100 \div 25$ = ______________
2. $330 \div 16$ = ______________
3. $407 \div 37$ = ______________
4. $562 \div 84$ = ______________
5. $646 \div 71$ = ______________
6. $950 \div 100$ = ______________
7. $1,000 \div 200$ = ______________
8. $1,200 \div 36$ = ______________
9. $1,540 \div 50$ = ______________
10. $2,003 \div 66$ = ______________

___   ___ ___ ___ ___ ___ ___ ___ ___ ___
7   5 9 2 10 4 1 8 6 3

What Number Am I?

Name ______________________________ Date ______________

1 I am a 2-digit number. The sum of my digits is 11. I am divisible by both 4 and 7.

What number am I? ____________________

2 I am a 2-digit number divisible by 4, 6, and 7.

What number am I? ____________________

3 I am a 2-digit number divisible by 19. The sum of my digits is 14.

What number am I? ____________________

4 I am a 3-digit number divisible by 7, but not 2. The sum of my digits is 4.

What number am I? ____________________

5 I am a 3-digit number less than 300. I am divisible by 2 and 5, but not 3. The sum of my digits is 7.

What number am I? ____________________

6 I am a 3-digit number divisible by 3. My tens digit is 3 times as great as my hundreds digit, and the sum of my digits is 15. If you reverse my digits, I am divisible by 6, as well as by 3.

What number am I? ____________________

The Squirm-ulator

Name ______________________________ Date ______________

Figure It Out!

1. Help out Squirmy Worm. What do you get when you multiply 6 by 7, then subtract 13? Use a calculator to check the answer.

____

2. Squirmy multiplies 8 by 5, then divides the product by 4. What is the answer?

____

3. Moovis the Cow multiplies 11 by 14. Then she divides the product by 7. What is the answer? ______________________________

4. Multiply the number of days there are in a week by 12. Subtract 24. What is the answer?

____

5. How old are you? Multiply your age in years by 17. Then add or subtract to get a total of 200. What number did you add or subtract?______________________________

SUPER CHALLENGE: On which day of the month were you born? Multiply this number by 3. Is the product higher than 100?

Number Stumper

Name ______________________ Date ______________

Put ÷, x, +, or – in the boxes to make correct math sentences.

1.

12		5		2	=9
2		6		4	=8
3		9		3	=4
=8		21		=11	

2.

8		5		4	=17
6		6		8	=28
15		21		9	=4
=29		=32		=23	

3.

15		2		10	=3
5		12		13	=30
4		3		9	=3
=12		=8		=14	

4.

11		9		4	=5
2		7		4	=10
3		8		2	=12
=19		=2		=18	

Magnetic Math

Name ______________________________ Date ______________

Riddle: What did one magnet say to the other magnet?

Decoder

1/3	T
3/165	V
1/32	A
15/20	B
21/67	F
99/312	T
61/83	K
59/83	U
4/156	L
3/8	C
11/121	W
3/11	A
22/121	T
2/32	I
4/11	U
14/20	R
3/12	N
20/67	M
3/156	E

Do each subtraction problem. Then use the Decoder to solve the riddle by filling in the spaces at the bottom of the page.

1. 2/3 – 1/3 = ________
2. 5/8 – 2/8 = ________
3. 7/11 – 4/11 = ________
4. 19/20 – 5/20 = ________
5. 27/32 – 20/32 – 6/32 = ________
6. 42/67 – 18/67 – 4/67 = ________
7. 79/83 – 11/83 – 9/83 = ________
8. 100/121 – 78/121 = ________
9. 44/156 – 29/156 – 12/156 = ______
10. 247/312 – 59/312 – 39/312 – 50/312 = ______

"YO ___ ___ ___ ___ ___ ___ ___ ___ ___ ___."

7 5 1 10 4 3 2 8 6 9

Name ____________________ Date __________

Solve the problems. Rename the answers in lowest terms. Then connect the dot beside each problem to the dot beside its answer. One line has been drawn for you.

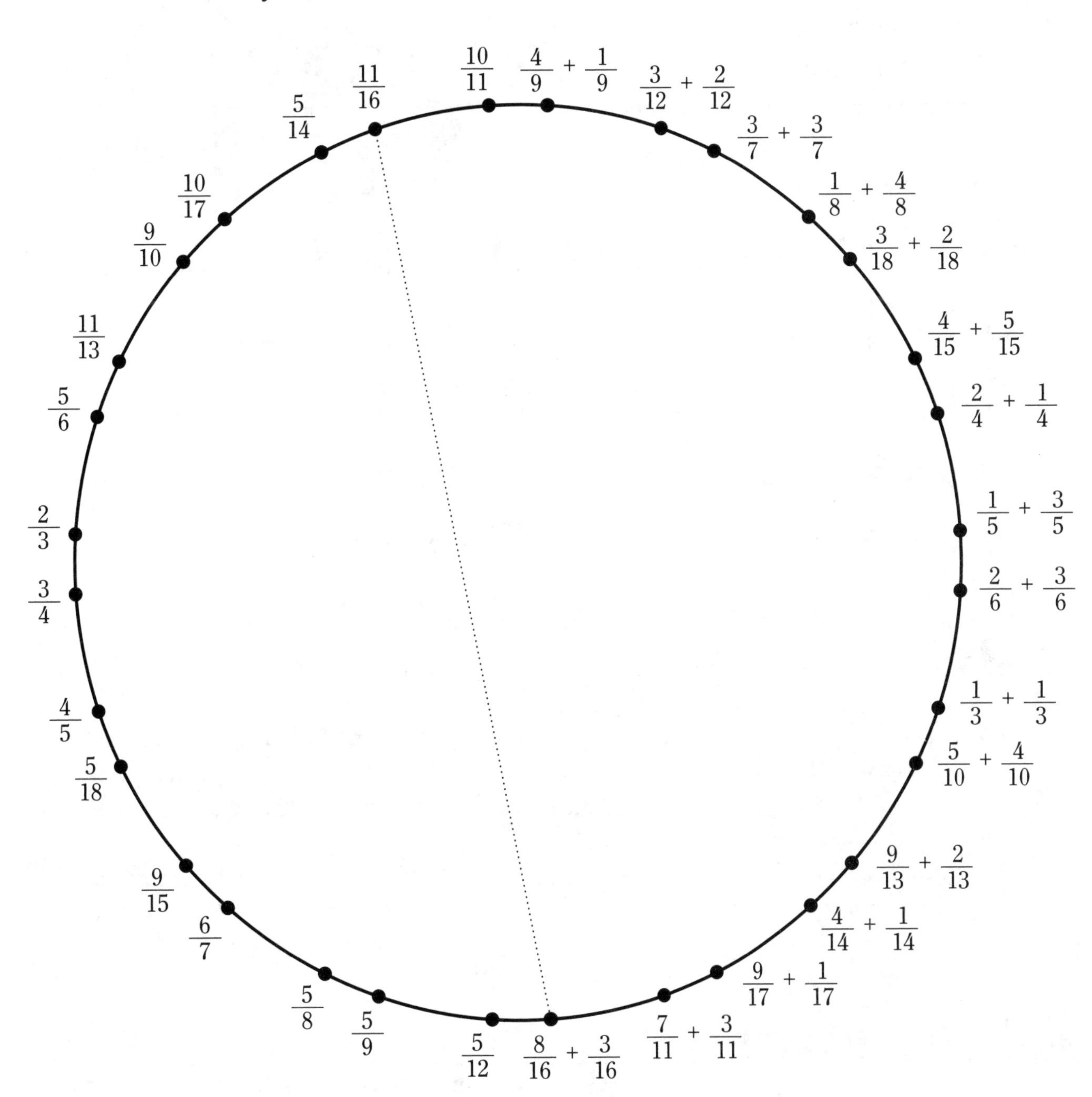

Everyone Needs Math!

Name ______________________________ Date ______________

Riddle: Why did the artist need math?

To find the answer to the riddle, solve the multiplication problems. Then, match each product with a letter in the Key below. Write the correct letters on the blanks below.

1. $3 \times \frac{1}{2} =$ ________
2. $5 \times \frac{1}{3} =$ ________
3. $2 \times \frac{1}{6} =$ ________
4. $4 \times \frac{2}{5} =$ ________
5. $3 \times \frac{3}{4} =$ ________
6. $2 \times \frac{7}{8} =$ ________
7. $6 \times \frac{6}{9} =$ ________
8. $5 \times \frac{2}{3} =$ ________
9. $4 \times \frac{4}{7} =$ ________
10. $6 \times \frac{9}{11} =$ ________

Key

3/2	M	45/11	F	10/3	E
16/7	Y	9/4	D	8/7	G
6/3	W	2/3	Z	36/9	R
2/6	N	54/11	U	8/5	S
14/8	B	3/6	T	5/3	B

Riddle Answer: **HE PAINTE** ___ ___ ___ ___ ___ ___ ___ ___ ___ ___.

5 2 9 3 10 1 6 8 7 4

Spring Flowers

Name ______________________________ Date ______________

Rename the fractions. If the fraction equals $\frac{1}{2}$, color the shape orange. If the fraction equals $\frac{1}{3}$, color the shape yellow. If the fraction equals $\frac{1}{4}$, color the shape blue. Finish the design by coloring the other shapes with the colors of your choice.

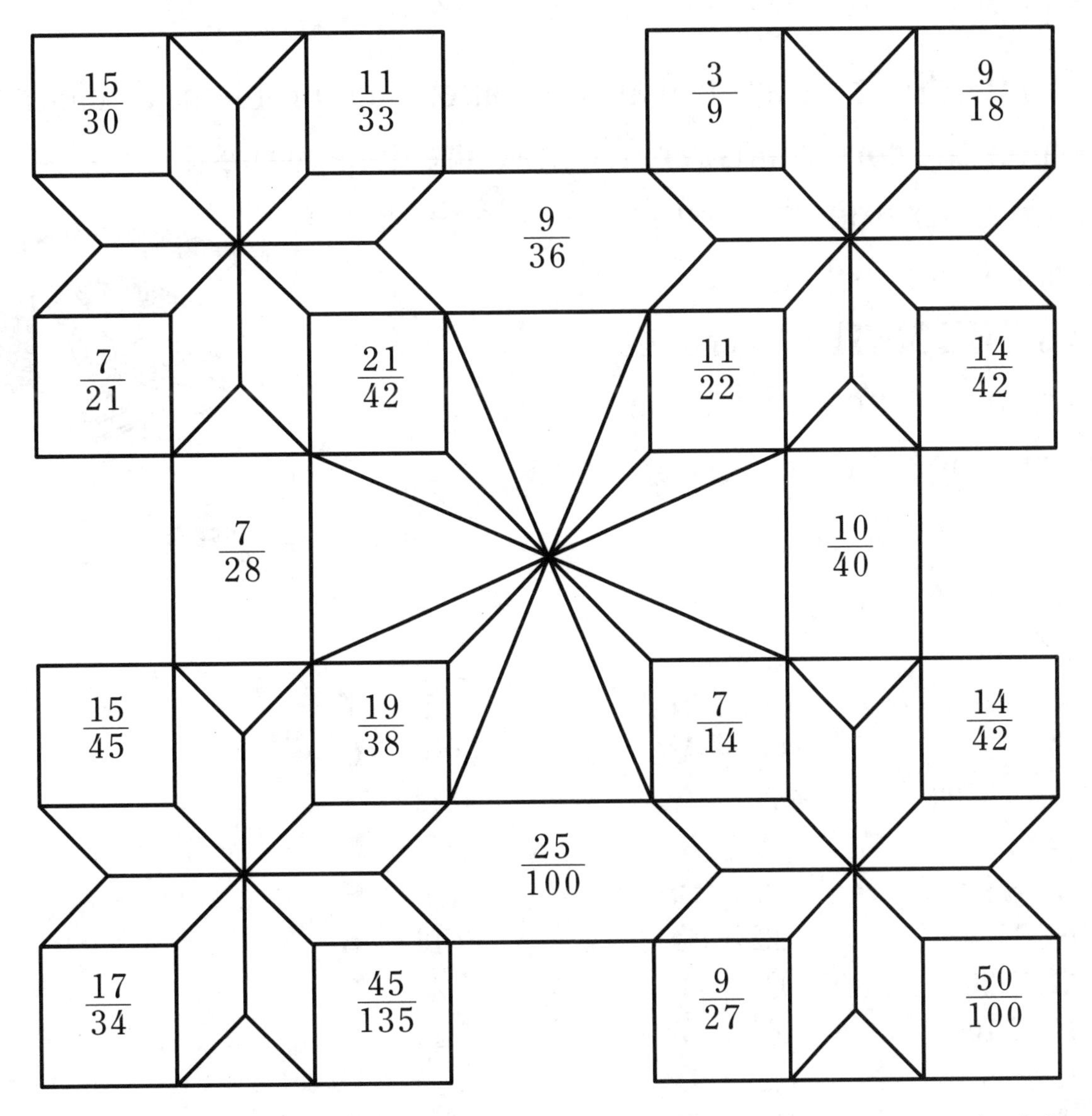

Taking It Further: Complete the squares so that each box adds up to 1. Use the following fractions once: $\frac{8}{14}$, $\frac{4}{14}$, and $\frac{1}{14}$.

a.

$\frac{1}{7}$	$\frac{1}{7}$
$\frac{6}{14}$	

b.

$\frac{1}{2}$	$\frac{2}{14}$
	$\frac{4}{14}$

c.

	$\frac{2}{14}$
$\frac{1}{7}$	$\frac{1}{7}$

Fruity Fractions

Name ______________________________________ Date ______________

✏ Why does a banana use suntan lotion? This question is a tricky one. So don't slip up! One way to find the answer is by turning these fractions into equivalent decimals.

DIRECTIONS: There are two answers after each problem. Circle the letter after the correct answer. When you're done, write the circled letters in order from the first problem to the last in the blank spaces below.

DOING THE MATH: To change a fraction to a decimal, divide the numerator by the denominator.

Example: 9/5 = 9 ÷ 5

$5\overline{)9} = 1.8$

A. 6/10	0.6	**S**	0.1	**T**
B. 4/9	3.2	**L**	0.4	**O**
C. 42/100	4.20	**A**	0.42	**I**
D. 13/5	2.6	**T**	5.3	**M**
E. 8/3	2.6	**W**	7.4	**B**
F. 11/50	0.22	**O**	2.12	**E**
G. 5/20	.025	**I**	0.25	**N**
H. 7/100	0.07	**T**	7.10	**B**
I. 16/5	6.2	**D**	3.2	**P**
J. 3/4	5.7	**U**	0.75	**E**
K. 14/3	4.6	**E**	9.3	**A**
L. 8/1000	0.008	**L**	.008	**R**

Why does a banana use suntan lotion?

___ ___   ___ ___   ___ ___ ___ ’ ___   ___ ___ ___ ___ .

Kaleidoscope of Flowers

Name ______________________________ Date ______________

If the number has a 5 in the ones place, color the shape green.
If the number has a 5 in the tenths place, color the shape pink.
If the number has a 5 in the hundredths place, color the shape yellow.
Finish the design by coloring the other shapes with colors of your choice.

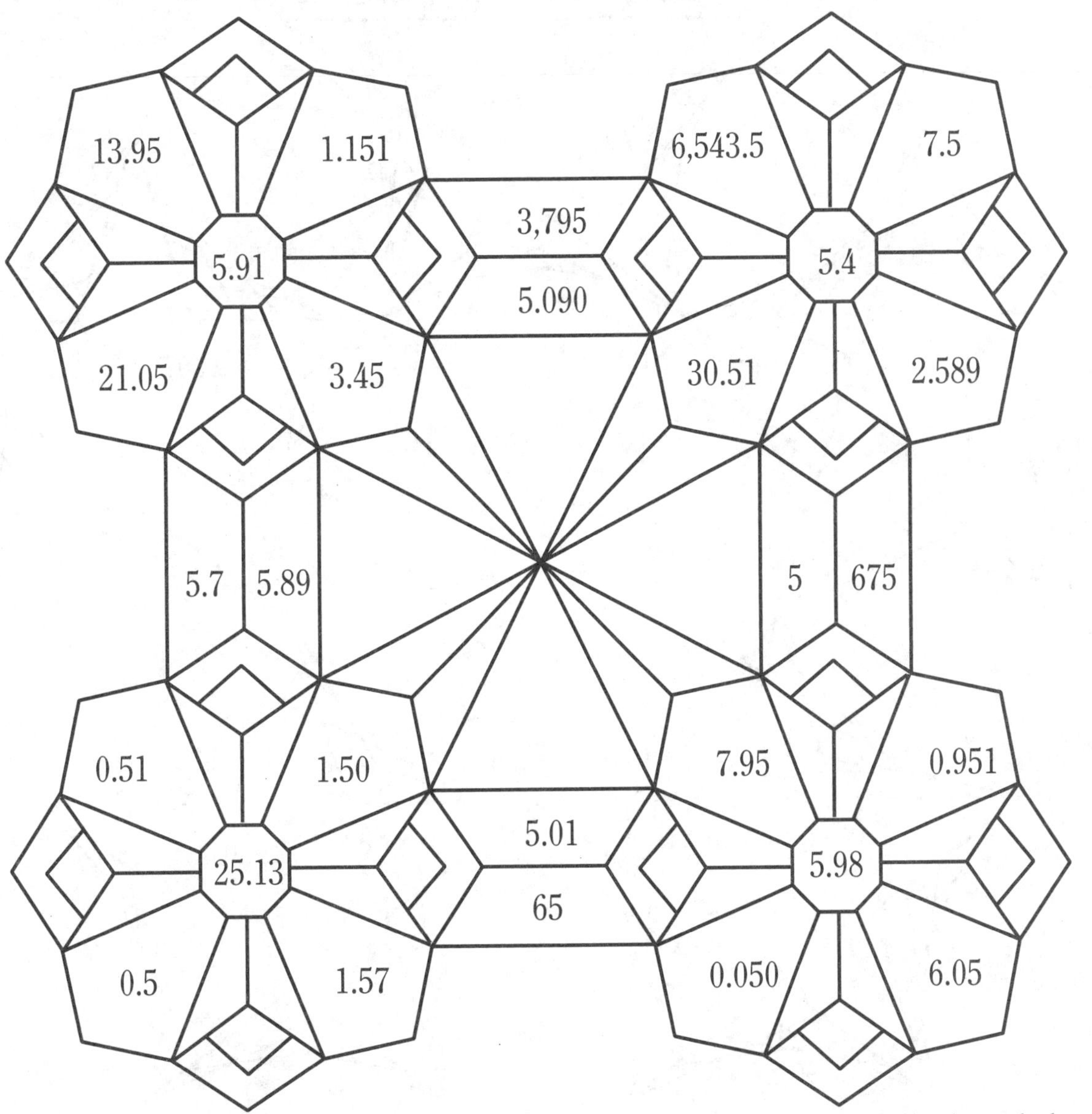

Taking It Further: Place the following decimals in the correct places on the lines below the dots: 4.9, 1.7, 2.5, and 0.2.

Ring of Stars

Name ______________________________ Date ______________

Form a star polygon by connecting the dots beside the decimals inside the pentagon. Begin with the smallest decimal and continue connecting the dots until you reach the largest decimal. The first and last lines have been drawn for you.

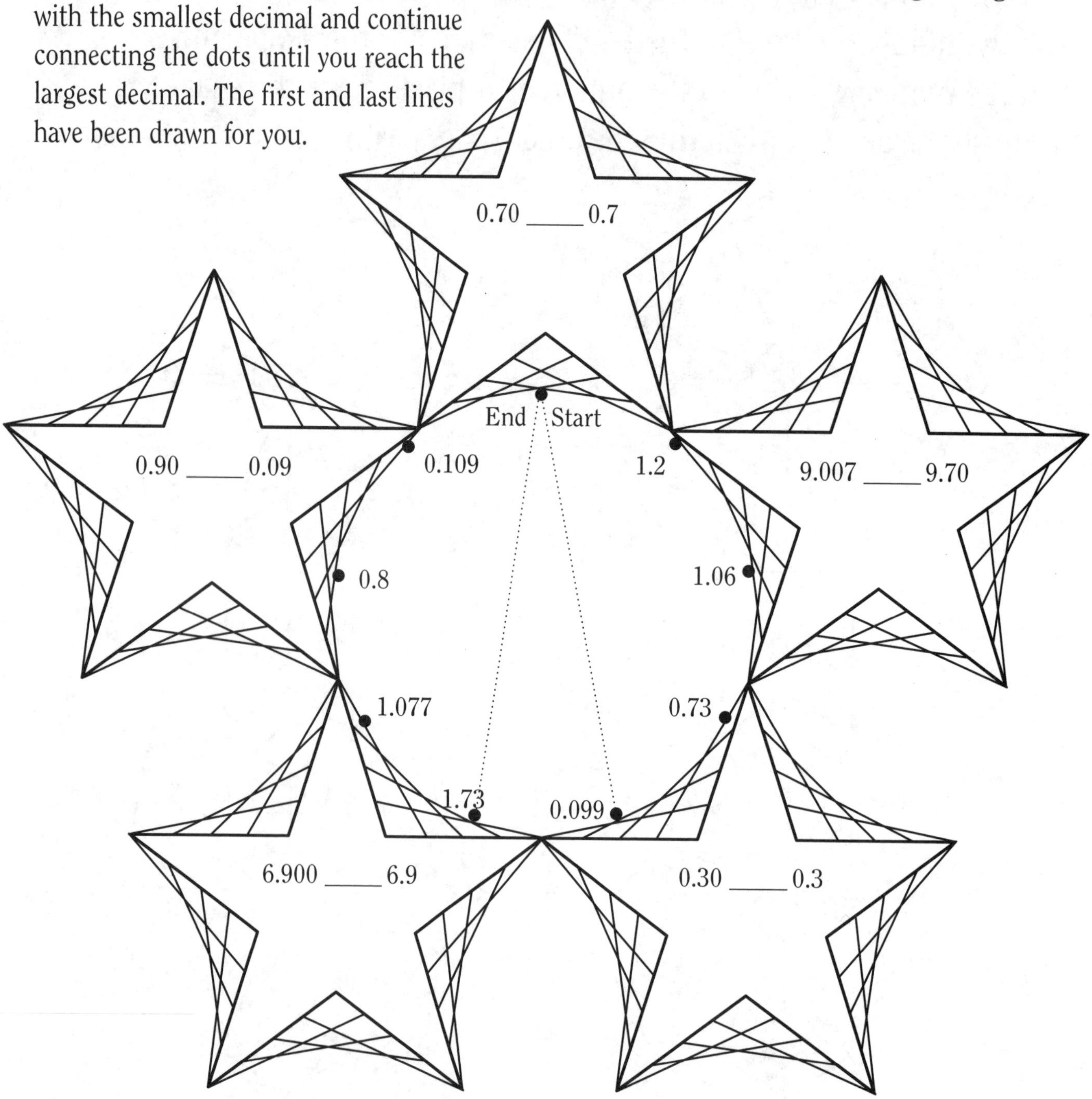

0.70 _____ 0.7

0.90 _____ 0.09

9.007 _____ 9.70

End Start

0.109

1.2

0.8

1.06

1.077

0.73

1.73

0.099

6.900 _____ 6.9

0.30 _____ 0.3

Taking It Further: Compare each pair of decimals inside the outer stars by writing <, >, or = on each blank line. If the decimals inside the star are equivalent, color the star blue. If the decimals inside the star are not equivalent, color the star red. Finish the design by coloring the rest of the shapes with the colors of your choice.

Dottie's Quilt

Name ______________________________ Date ______________

Solve the problems. ◆ If the answer is 100 or greater, color the shape pink. ◆ If the answer is less than 100, color the shape green. ◆ Finish the design by coloring the other shapes with the colors of your choice.

Taking It Further: Rewrite this problem on another sheet of paper and solve it. 2.99 + 14.1 + 787.02 + 16 = ______________

Lantern Glow

Name ______________________________ Date ____________

Solve the problems. ◆ If the number in the tenths place is 0, 1, 2, or 3, color the shape green. ◆ If the number in the tenths place is 4 or 5, color the shape red. ◆ If the number in the tenths place is 6, 7, 8, or 9, color the shape pink. ◆ Finish the design by coloring the other shapes with the colors of your choice.

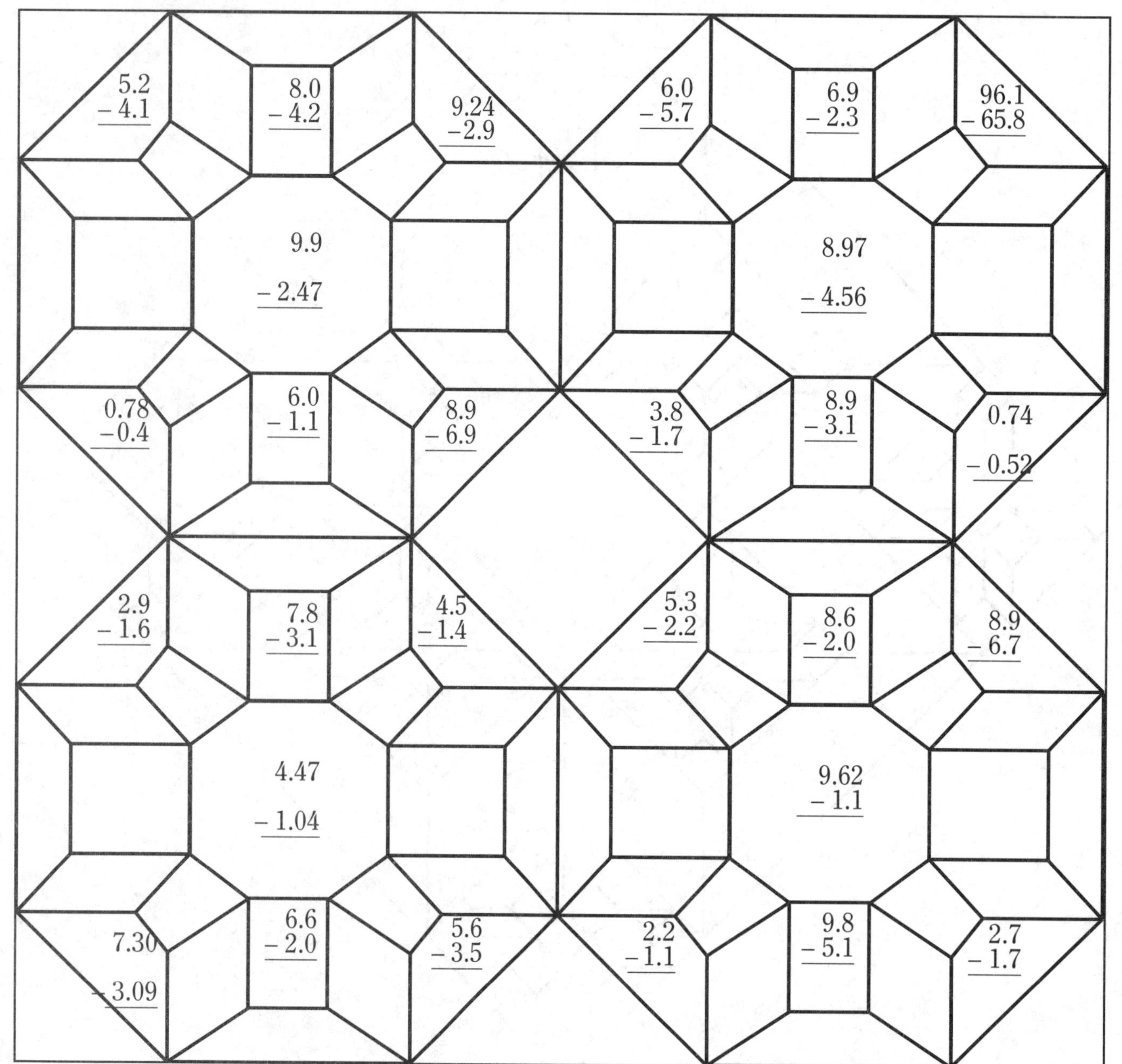

Taking it Further: Rewrite these problems on another sheet of paper and solve them.

a. 3.4 – 1.009 = _____ b. 79.03 – 9.4 = _____ c. 81.02 – 4.99 = _____

d. 7.9 – 4.012 = _____

Home Improvement?

Name ________________________________ Date ______________

✏ Michelle's family just bought a new house. Workers were putting a few last minute touches on it before the family moved in. But the day turned into one big disaster! Michelle will tell you all about it.

DIRECTIONS: To complete Michelle's story, solve the problem next to each worker's name. ◆ Next, find your answer below a blank in the story. ◆ Write that worker's name in the blank. ◆ When you're done, read Michelle's story.

WORKER'S NAMES

1. 5% of 60 = _____ Paul Plumber
2. 50% of 1000 = _____ Robert Roofer
3. 6% of 450 = _____ Penny Painter
4. 8% of 90 = _____ Alan Architect
5. 40% of 200 = _____ Gilbert Gardener
6. 30% of 620 = _____ Elway Electrician
7. 20% of 100 = _____ Carlton Carpenter

MICHELLE'S STORY

I'll never forget the day the workers showed up at our new house! First, ____________ (186) dropped his screwdriver on the floor. Then ____________ (7.2) slipped on it and accidentally knocked a can of paint onto ____________ (20)'s diagrams. He was pretty upset about it and asked ____________ (3) to drive him to pick up new ones. While they were pulling out of the driveway, they ran over ________ (500)'s tools. ____________ (80) yelled for them to stop but they didn't hear him. ____________ (27) looked at all of this in disbelief. And so did I!

The Next Number . . .

Name ______________________________ Date ______________

✏ Sometimes sets of numbers have something in common. They follow a pattern. Take a look at the numbers 4, 6, 8, and 10. As the pattern continues, each number gets larger by 2. Try completing the number patterns in the problems below. Some are tougher to figure out than others. Give 'em a try. Good luck! Use the space below and to the right to work out the problems.

1. 8, 11, 14, 17, 20, _____, _____, _____

2. 27, 29, 31, 33, 35, _____, _____, _____

3. 2, 7, 12, 17, 22, 27, _____, _____, _____

4. 5, 9, 14, 23, 37, 60, _____, _____, _____

5. 39, 46, 53, 60, 67, 74, _____, _____, _____

6. 6, 7, 13, 20, 33, 55, _____, _____

7. 4, 15, 26, 37, 48, _____, _____, _____

8. 93, 116, 209, 325, 534, 859, _____, _____

Come up with several number patterns of your own. Ask someone to complete the pattern.

Times Terms

Name ______________________________ Date ______________

Write the multiplication word that fits each clue in the box. When you finish, copy the letters in the shaded boxes. Unscramble these letters to form another multiplication word.

1. Any number multiplied by this number comes out 0.

2. Another word for multiplied by is ______.

3. This is one of the numbers you multiply.

4. Multiply a number by 3, and you ______ that number.

5. Multiply a number by 2 to get the same answer as adding a ______.

6. The answer when you multiply is called the ______.

7. Its math symbol is .

8. Multiplication is the same as repeated ______.

9. You can multiply if you have groups that are the ______ ______ (2 words).

Write the letters from the shaded boxes here.

Now unscramble them to make another word.

Tell what this word means. ______________________________

____

____

8 x 3 = 24

6 x 7 = 42

Mixed Operations

1. One-half of a number added to one-fourth of 96 is 30.

   What is the number? ______________________

2. If you triple a number you will have one-half the number of hours in two days.

   What is the number? ______________________

3. If you double a number, you will get the same as the triple of one-fourth of 24.

   What is the number? ______________________

4. If you subtract a number from the square of 7 you will get one-fourth the product of 9 and 8.

   What is the number? ______________________

5. One-fifth of a number, subtracted from 20, is the same as one-fourth of 32.

   What is the number? ______________________

6. Think of two numbers whose greatest common factor is 12. If you divide the lesser of the two numbers by that greatest common factor, you get one-sixteenth of the other number.

   What are the numbers? ______________________

What a Sale!

Name ______________________________ Date ______________

✏ There's a big sale over at the Clothing Coop. Ashley and Deondra are there to buy a few things. "How will we know how much money we're saving on each item?" Deondra asked. "Say a jacket that costs $32.00 has a sale tag that says 20% off," Ashley explained. "That means the store will take $ .20 off each dollar. In other words, the store will take a total of $6.40 off the original price of the jacket."
Help the girls figure out how much money the store will take off the other items they want to buy.

DOING THE MATH:
20 PERCENT OFF $32.00

Multiply the same way you would with whole numbers.

```
  $32.00
x $  .20
  64000
```

Add the number of decimal places.

```
  $32.00
x $  .20
  64000
```

4 decimal places altogether.

Move the decimal point 4 places to the left.

```
  $32.00
x $  .20
  6.4000
```

ANSWER
$6.40

1. a. What amount should be taken off the original price? __________

b. What price will the girls pay for the pants? __________

$24.00 30% off

2. a. How much should be taken off the original price? __________

b. What will they pay for the blouse? __________

$17.00 40% off

3. a. What amount should be taken off the original price? __________

b. What's the sale price of the pocketbook? __________

$22.00 15% off

4. a. How much should they take off the original price? __________

b. What's the sale price of the shoes? __________

$65.00 20% off

Multiplying & Dividing

Name ______________________ Date ____________

Choose one number from the triangle and one from the circle to answer each question.

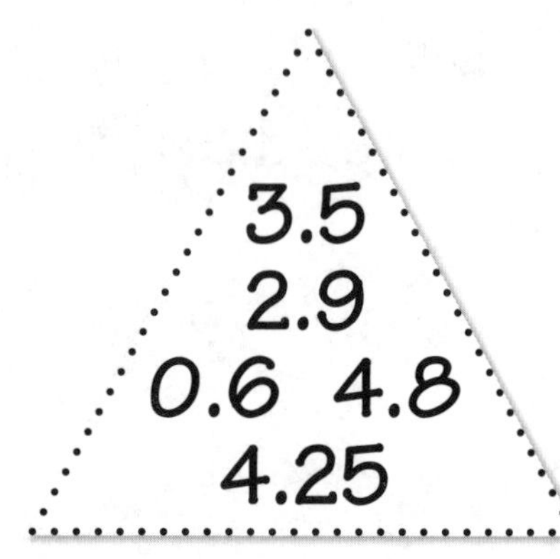

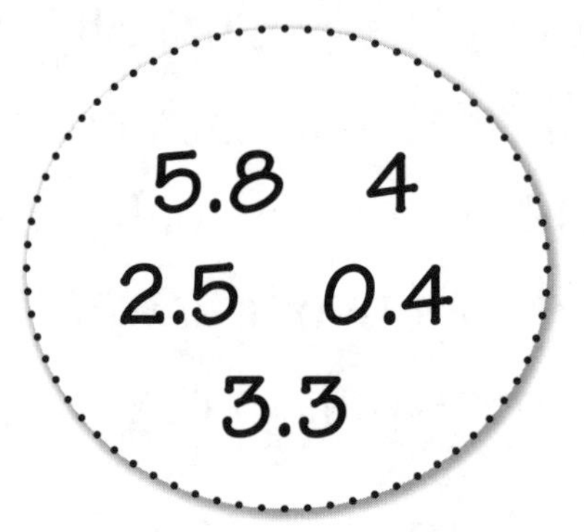

Two numbers have a product of 8.75.

What are the numbers? ______________________

Two numbers have a product of 17.

What are the numbers? ______________________

Two numbers have a product that is less than 1.

What are the numbers? ______________________

Two numbers have a product that is greater than 25.

What are the numbers? ______________________

Changing Shapes

Name ______________________________ Date ______________

Riddle: How did the square become a triangle?

To find the answer to the riddle, solve the multiplication problems here. (Don't forget units.) Then,

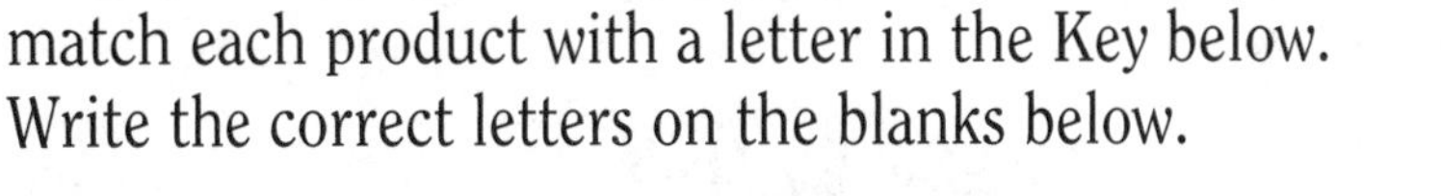

match each product with a letter in the Key below. Write the correct letters on the blanks below.

1. Joe has 2 apples. Tim has 2 times as many apples as Joe has. How many apples does Tim have? _______
2. Kendra has 3 books. Paula has 3 times as many books as Kendra has. How many books does Paula have? _______
3. Cliff has 5 times as many baseball caps as Wayne has. Wayne has 5 baseball caps. How many baseball caps does Cliff have? _______
4. Jorge has 10 oranges. Wendy has 2 times as many oranges as Jorge has. How many oranges does Wendy have? _______
5. Martha has 6 times as many coats as Russell has. Russell has 5 coats. How many coats does Martha have? _______
6. Debbie has 9 pairs of shoes. How many shoes does she have in all? _______
7. Michael has 8 bunches of bananas. Each bunch has 7 bananas. How many bananas does he have in all? _______
8. Leroy has 11 times as many pencils as Renee has. Renee has 11 pencils. How many pencils does Leroy have? _______
9. Steve has 6 video games. Jack has 8 times as many video games as Steve has. How many video games do Steve and Jack have in all? _______
10. Carla has 7 chairs. Kim has 7 times as many chairs as Carla has. How many more chairs does Kim have than Carla? _______

Key

4 apples	T
20 oranges	C
18 shoes	N
56 bananas	C
111 pencils	I
54 video games	E
48 video games	F
30 coats	U
2 apples	S
42 chairs	A
15 bananas	K
9 books	R
25 caps	R
121 pencils	O
40 coats	B

Riddle Answer: IT ___ ___ ___ ___ ___ ___ ___ ___ ___ ___.

7 5 1 10 4 8 3 6 9 2

Weatherman

Name ______________________________ Date ______________

Figure It Out!

1. Showers on Monday morning produced 0.5 inches of rain by noon. By 6 p.m., a total of 2 inches of rain had fallen. How many inches of rain fell between noon and 6 p.m.? __________

2. On Tuesday, 1.2 inches of rain fell. Two more inches of rain fell the next day. How many inches of rain fell on Wednesday? __________________

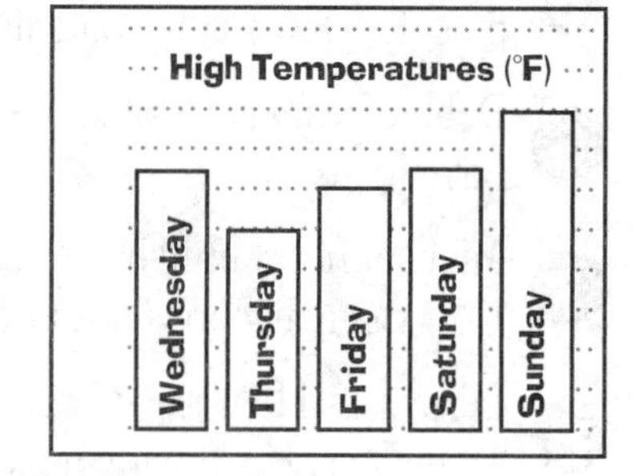

3. The graph shows the high temperatures for Wednesday through Sunday. On which day was the highest temperature reached? The lowest? What was the difference between the two temperatures? ______________________________

4. Between which two days did the temperature drop 15 degrees? Increase by 15 degrees? ______________________________

5. Saturday's low temperature was 38°. How many degrees did the temperature rise to reach Saturday's high temperature? ______________________________

SUPER CHALLENGE: What was the average high temperature for all five days shown on the graph?

Volume Pops Up Everywhere!

Name ______________________________ Date ______________

Look around the room. Do you see any of the shapes shown here?

Cylinder Cube Cone

These shapes are three-dimensional. That means that they are solid—you can touch them with your hands. (You can't hold a two-dimensional shape like a circle, square, or triangle.) We measure three-dimensional shapes in a special way—using volume. Volume tells how much the shape can hold inside.

Ready to learn about volume? Let's go!

You Need:
2-lb bag of unpopped popcorn
ice cream cone
empty drink box with top cut off
empty 8-oz yogurt cup
8- or 9-inch pie plate

1. Start with the cone and the yogurt cup. How many cones do you think it will take to fill the cup with popcorn?

   ____________ cones

   Fill the cone with popcorn. Then pour it into the cup. Keep filling the cup until you think it's half filled.

   Do you want to change your guess?

   New guess: ____________ cones

   Now finish filling the cup. How many cones did it take?

   ____________ cones

2. How many cups of popcorn do you think it will take to fill the pie plate? Start pouring popcorn from the cup to the pie plate. When you think the pie plate is half filled, guess again. Then fill it all the way. How many cups did it take?

   ____________ cups

3. Which do you think holds more, the cup or the drink box? How could you find out? Test your ideas. Which holds more?

   ____

4. How many drink boxes do you think it would take to fill the pie plate? Try it.

   ____________ drink boxes

Now pop the popcorn, fill the cone with ice cream, and have a volume party!

Brain Power
Try more volume experiments with other containers.

Get an "Angle" on Inventions

Name ______________________ Date __________

✏ Everything that people use in their daily lives was invented by someone—things like the ironing board, the cash register, and ear muffs. In this activity, we ask you to match inventions such as these to their inventor. Follow the directions below to get a new "angle" on a few famous inventions.

40°

90°

180°

DIRECTIONS:

- Take a look at the angle that appears before each statement.
- Estimate the measure of the angle in degrees using the 40°, 90°, and 180° angles as a guide.
- Next, circle the name of the invention that appears next to the best estimate of that angle.
- Write the correct invention in the space provided in the statement.

1. The __________ was invented in 1888 by A.B. Blackburn.

2. S. Boone invented the __________ in 1892.

3. The __________ was invented in 1912 by Garrett A. Morgan.

4. The __________ was invented in 1879 by James Ritty.

5. In 1877, Chester Greenwood invented __________.

6. In 1935, Laszlo and Georg Biro established themselves as the first inventors of the __________.

7. In 1902, the __________ was invented by Miller Hutchison.

8. Other inventors expanded on her invention in later years. But Mary Anderson was the inventor of the first __________ in 1903.

11° hearing aid **90° ironing board** **130° windshield wiper**
160° cash register **80° ear muffs** **175° railway signal**
20° ballpoint pen **110° gas mask**

Break the Ice With Perimeter and Area

Name ______________________________ Date ______________

Jessie is building ice skating rinks for her friends. She measures the size of each rink in two ways—**perimeter** and **area**. Perimeter tells the measurement **around** the rink. Area tells how many square units fit **inside** each rink. Some rinks have the same area but different perimeters. Try some building yourself!

You Need:
square crackers or square counters

Here's the rink Jesse built for Shawn. Its area is 4. Its perimeter is 8.

What to Do:

Use the square crackers to help you answer the questions. Then draw how the crackers look.

1. Shawn wants a bigger rink. He wants it to have a perimeter of 12 and an area of 8. What can you add to Shawn's rink? Draw what it will look like.

2. Gil also wants a rink with a perimeter of 12. But he wants it to be square. What will it look like? What will its area be? Draw what it will look like.

3. The area of Rita's rink is 12. Its perimeter is 14. What does her rink look like? Draw it.

4. Sonia wants her rink to have an area of 16. She says it can be shaped like a square or a rectangle. What could the rink look like? What will its perimeter be? Draw it.

5. José wants a rink with an area of 24. It can be any shape. What are some of the shapes it could be? What are their perimeters? Draw one example.

Brain Power
Draw a shape whose perimeter and area are the same number.

Check This Out!

Name ______________________________ Date ______________

Riddle: Why did the sick book visit the library?

To find the answer to the riddle, solve the multiplication problems. Then, match each product with a letter in the Key below. Write the correct letters on the blanks below.

1. How many inches are there in 1 foot? ____
2. How many inches are there in 2 feet? ____
3. How many inches are there in 4 feet? ____
4. How many inches are there in 5 feet? ____
5. How many inches are there in 7 feet? ____
6. How many inches are there in 9 feet? ____
7. How many inches are there in 10 feet? ____
8. How many inches are there in 6 feet? ____
9. How many inches are there in 12 feet? ____
10. How many inches are there in 15 feet? ____

Key

60	E	72	D	88	M
84	C	140	I	64	P
180	O	120	H	12	U
100	G	110	L	24	K
108	T	144	E	48	C

Riddle Answer: **TO GET** "___ ___ ___ ___ ___ ___ ___   ___ ___ ___"

3 7 4 5 2 9 8   10 1 6

A Royal Riddle

Name ______________________ Date ____________

Riddle: Where does a king stay when he goes to the beach?

To find the answer to the riddle, solve the multiplication problems. Then, match each product with a letter in the Key below. Write the correct letters on the blanks below.

1. How many minutes are there in 1 hour? ____
2. How many minutes are there in 2 hours? ____
3. How many minutes are there in 4 hours? ____
4. How many minutes are there in 5 hours? ____
5. How many minutes are there in 7 hours? ____
6. How many minutes are there in 10 hours? ____
7. How many minutes are there in 11 hours? ____
8. How many minutes are there in 15 hours? ____
9. How many minutes are there in 18 hours? ____
10. How many minutes are there in 20 hours? ____

Key

600	S	1,240	M	900	E
420	C	120	D	450	B
1,200	N	180	X	1,100	I
660	S	300	A	240	A
1,080	T	60	L	360	O

Riddle Answer: A ___ ___ ___ ___ ___ ___ ___ ___ ___ ___

6 3 10 2 5 4 7 9 1 8

Time Travels

Name ______________________________ Date ______________

Riddle: What part of a cowboy is the saddest?

Find the answer. Then use the Decoder to solve the riddle by filling in the blanks at the bottom of the page.

1. Time in New York is one hour later than in Chicago. If it's 10 a.m. in New York, what time is it in Chicago? ___
2. If it's 4 p.m. in New York, what time is it in Chicago? ___
3. If it's 8:30 p.m. in Chicago, what time is it in New York? ___
4. Time in New York is three hours later than time in Los Angeles. If it's 10 a.m. in New York, what time is it in Los Angeles? ___
5. If it's 2:30 p.m. in New York, what time is it in Los Angeles? ___
6. If it's 5:20 a.m. in Los Angeles, what time is it in New York? ___
7. When it is 11:17 a.m. in New York, it is 8:17 a.m. in Los Angeles. When it's 11:17 a.m. in Los Angeles, what time is it in New York? ___
8. Time in Los Angeles is two hours earlier than time in Chicago. If it's 8 p.m. in Los Angeles, what time is it in Chicago? ___
9. When it's 6 p.m. in Los Angeles, it's 9 p.m. in New York. What time is it in Chicago? ___
10. When it's 3:11 p.m. in Chicago, it's 4:11 p.m. in New York. What time is it in Los Angeles? ___

Decoder

8:20 a.m........ **E**
3:00 p.m........ **J**
2:20 a.m....... **K**
5:11 p.m....... **C**
11:30 a.m...... **L**
5:30 p.m......... **I**
8:00 p.m. ..... **N**
2:17 p.m....... **S**
1:11 p.m........ **E**
11:00 a.m. ... **M**
12:11 p.m..... **O**
9:00 a.m........ **U**
11:00p.m....... **T**
6:00 a.m. .... **W**
9:30 p.m. ...... **A**
7:30 p.m. ...... **Z**
10:00 p.m..... **B**
7:00 a.m....... **S**
2:00 p.m....... **G**

HI ___ (4) ___ (8) ___ (5) ___ (1) ___ (10) ___ (2) ___ (6) ___ (3) ___ (9) ___ (7)

A Sick Riddle

Name ______________________________ Date ______________

Riddle: Why do people with colds get plenty of exercise?

Find the answers. Then use the Decoder to solve the riddle by filling in the blanks at the bottom of the page.

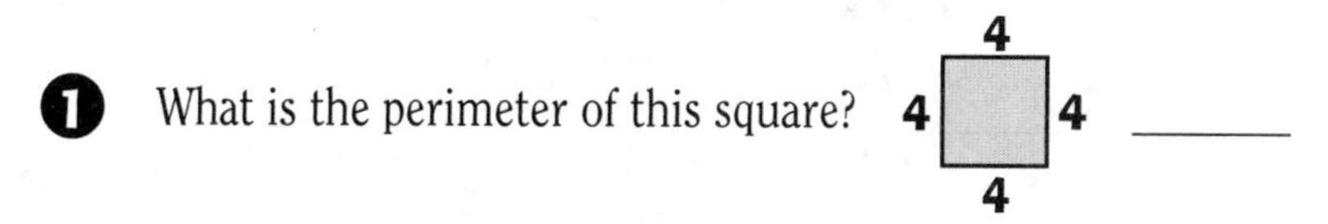

1. What is the perimeter of this square? (4, 4, 4, 4) ______

2. What is the perimeter of this rectangle? (8, 2, 8, 2) ______

3. What is the perimeter of this triangle? (3, 9, 7) ______

4. What is the perimeter of a square that is 10 inches long on one side? ______

5. A square's perimeter is 48 inches. How long is one side of the square? ______

6. A triangle with three equal sides has a perimeter of 27 inches. How long is one side of the triangle? ______

7. Each side of a pentagon is 11 inches long. What is the pentagon's perimeter? ______

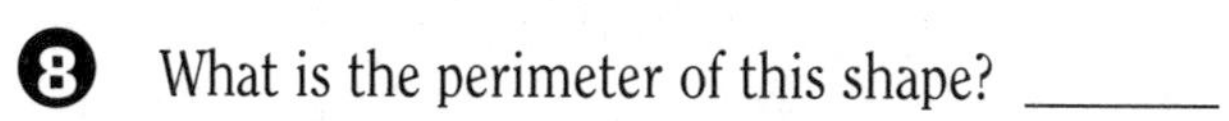

8. What is the perimeter of this shape? ______

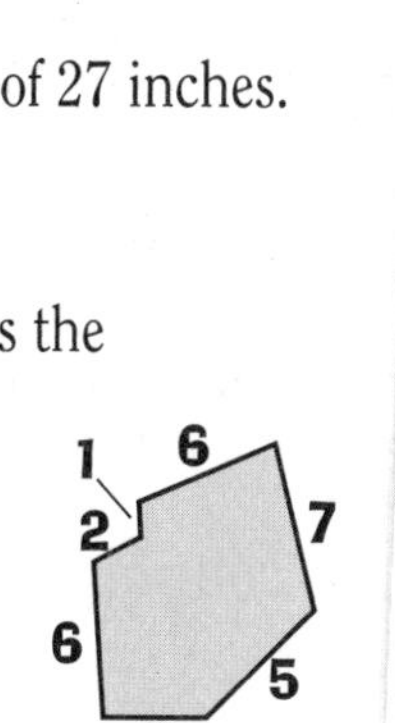

9. A magazine is 11 inches long and 8 inches wide. What is the magazine's perimeter? ______

10. A lawn is 23 feet long and 14 feet wide. What is the lawn's perimeter? ______

Decoder

19	O
74 inches	Q
30	N
25	A
38 inches	I
12 inches	S
40 feet	X
9 inches	N
15	B
74 feet	R
16	E
10 feet	D
20	R
22	A
32 inches	L
37 feet	M
40 inches	U
55 inches	S
15	C

THE ___ ___ ___ ___ ___ ___ ___ ___ ___ ___.

(9 2   8 3 5 1 7   10 4 6)

A Riddle to Dive Into

Name ______________________ Date ____________

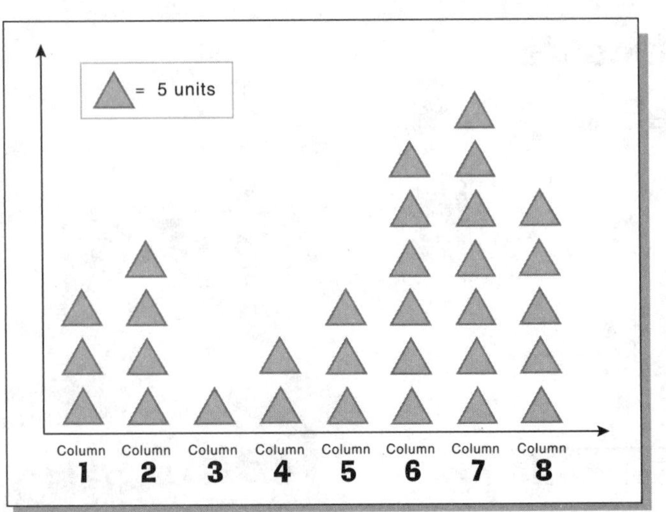

Riddle: How can you dive without getting wet?

Answer the questions about the graph. Then use the Decoder to solve the riddle by filling in the blanks at the bottom of the page.

1. How many units does one ▲ equal? _______
2. Which column has the most units? _______
3. Which column has the fewest units? _______
4. How many units are in column 2? _______
5. Which columns have the same number of units? _______
6. How many units are in column 8? _______
7. How many more units are in column 6 than in column 5? _______
8. How many fewer units are in column 3 than in column 7? _______
9. If the number of ▲s doubled in column 2, how many units would be in the column? _______
10. Which column has 1/3 of the units of column 6? _______

Decoder

40 units........... **G**
50 units........... **A**
column 7 .... **S**
column 1 .... **P**
column 4 .... **K**
30 units........... **N**
columns 2 and 8............ **E**
25 units........... **O**
20 units............. **I**
column 6 .... **H**
35 units.......... **W**
5 units............. **D**
15 units............ **V**
column 8 .... **T**
10 units............ **L**
column 3 .... **I**
columns 1 and 5............. **Y**

G ___ (6)   ___ (2) ___ (10) ___ (5) ___ (1) ___ (4) ___ (7) ___ (3) ___ (8) ___ (9).

How's Your Heart Rate?

Name ______________________ Date ____________

You Need:
◆ stopwatch or watch with a second hand ◆ tennis ball

Animals have hearts that do the same job as a person's heart. An animal's heart beats all day long to pump blood through its body. What's different about an animal heart and a human heart? The number of times it beats in a day.

Each day your heart beats about 100,000 times. That's enough times to pump almost 1,500 gallons of blood throughout your body! By the time you are 70 years old, your heart will have pumped about 38 million gallons of blood. No wonder it's important to keep your heart strong and healthy!

The number of times a heart beats in a certain amount of time is called **heart rate**. Check out the table to find some average animal heart rates. Then follow the steps to add your heart rate to the table.

ANIMAL	HEART RATE (for one minute)
Canary	1,000
Mouse	650
Chicken	200
Cat	110
Dog	80
Adult human	72
Giraffe	60
Tiger	45
Elephant	25
Gray whale	8
You	

How to Find Your Heart Rate

- Place two fingers on your neck or your wrist. Move them around until you feel a pulse beat.
- Count the beats for 30 seconds. Have a partner time you with the watch.
- Multiply the number of beats by two. That number is your heart rate for one minute.

Hearts Are Hard Workers

To prove it, try this. Squeeze a tennis ball as hard as you can and let go. That's how hard your heart works to pump blood through your body. Now try to squeeze the ball for one minute to match your heart rate. Not too easy, is it?

Answer these questions about animals' heart rates, using the information on the table.

1. Which animal's heart beats fastest in one minute? ____________

Which beats slowest? ____________

2. What do you notice about the size of the animal compared with its heart rate?

____

3. Where do you think a horse's heart rate might fit on the table? Explain your answer.

____

4. Which animal is your heart rate the closest to? ____________

Sampling Cereal

Name ________________________________ Date ______________

Are there more Ps than Qs in a box of letter-shaped cereal? How about the other letters? You could look at every piece of cereal in the box. But that could take a while. It might be dinnertime before you get to eat breakfast!

We've got a better idea. Take a sample. A sample is a small part of a larger group. Studying a sample can tell you a lot about the whole group. If you look at the letters in a small bowl of cereal, you can get a good idea about what's in the rest of the box. That leaves only one more thing to figure out: who gets to eat the last bowl!

You Need:
- box of alphabet cereal
- measuring cup
- pencil and paper

What to Do:

1. Measure out one cup of letter-shaped cereal. This is your sample.
2. Pick one piece of cereal out of the cup. Then make a mark on the tally sheet next to the correct letter.
3. Do this for all of the cereal in your sample cupful. Don't count broken pieces. (If you find more than one cereal of the same letter, just mark it again on your tally sheet like this: 𝍸 ||.)
4. Which letters have the most tally marks on your sheet? ______________________

____

Tally Sheet

A: ____________
B: ____________
C: ____________
D: ____________
E: ____________
F: ____________
G: ____________
H: ____________
I: ____________
J: ____________
K: ____________
L: ____________
M: ____________
N: ____________
O: ____________
P: ____________
Q: ____________
R: ____________
S: ____________
T: ____________
U: ____________
V: ____________
W: ____________
X: ____________
Y: ____________
Z: ____________

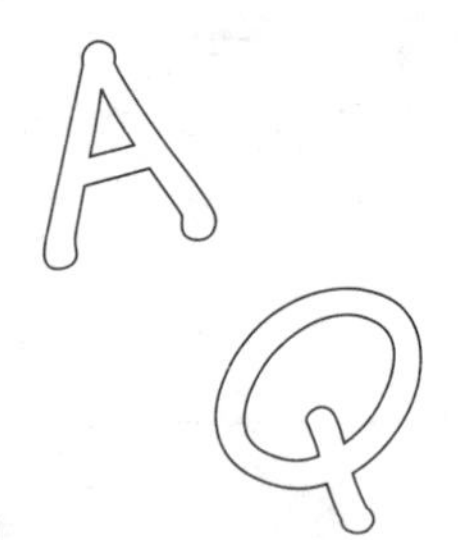

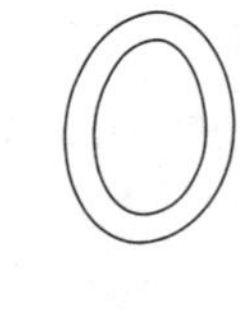

Brain Power

How could you use your sample to estimate how many of each letter are in the whole box?

Hot Dog–It's a Bar Graph!

Name ______________________ Date __________

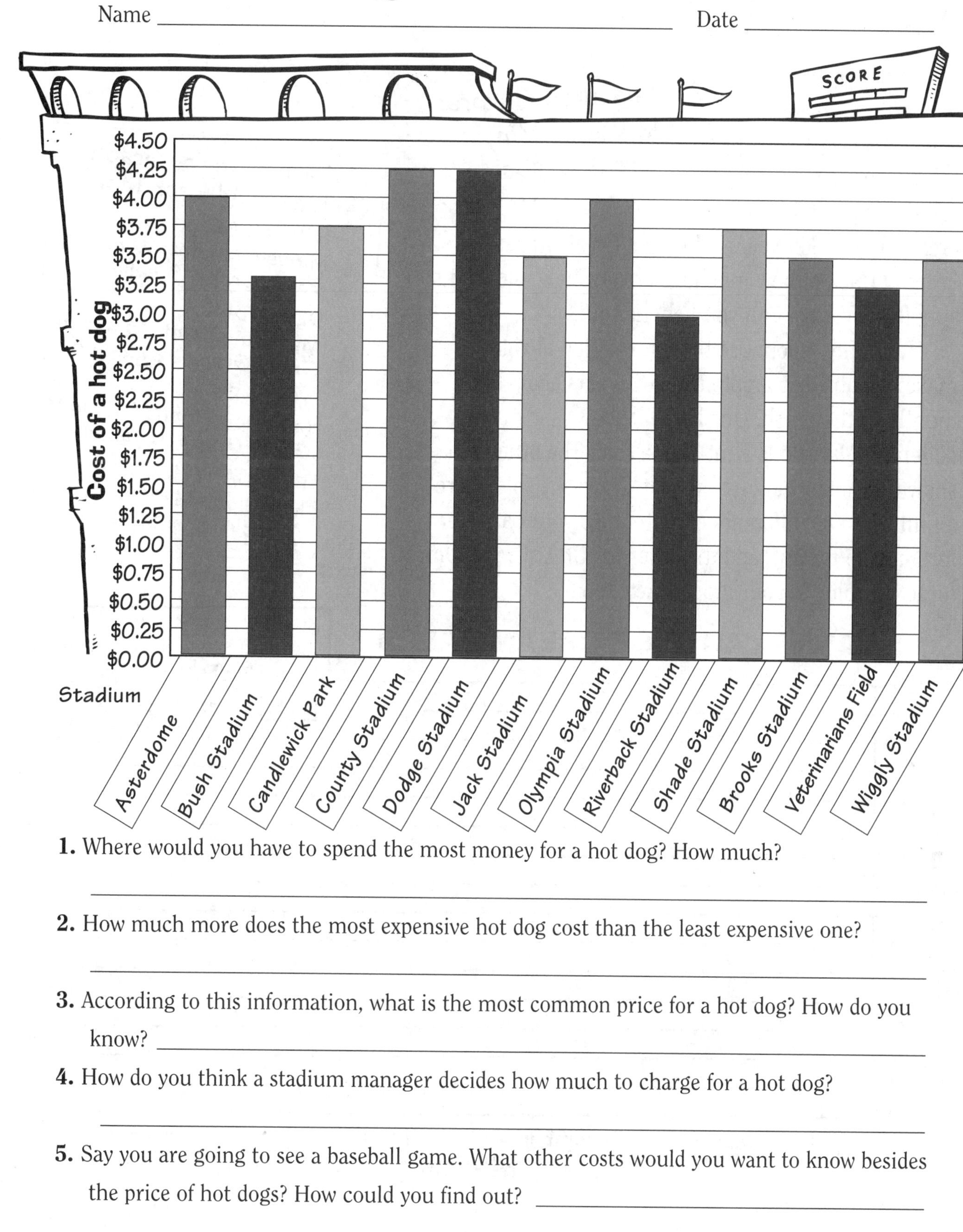

1. Where would you have to spend the most money for a hot dog? How much?

____

2. How much more does the most expensive hot dog cost than the least expensive one?

____

3. According to this information, what is the most common price for a hot dog? How do you know? ______________________

4. How do you think a stadium manager decides how much to charge for a hot dog?

____

5. Say you are going to see a baseball game. What other costs would you want to know besides the price of hot dogs? How could you find out? ______________________

We All Scream for Bar Graphs!

Name ______________________________ Date ____________

Our Featured Flavors:
Marvelous Mint!
Great Grape!

Vanna Lah owns her own ice cream shop. She keeps things simple by selling only two ice cream flavors: Marvelous Mint and Great Grape. But keeping track of how much ice cream she's sold isn't so simple for Vanna. Can you help her out?

Here's the scoop: Use the information at right to make a double bar graph. Draw two bars above each month on the graph. One bar will show how many gallons of Marvelous Mint Vanna sold. The other will show how many gallons of Great Grape she sold. Here's a hint: Use two different colors to draw your bars–one for Marvelous Mint, and the other for Great Grape. Don't forget to color in your graph's key, too.

Information for Graph:
Gallons Sold Each Month

NOVEMBER
Marvelous Mint: 10 gallons
Great Grape: 8 gallons

DECEMBER
Marvelous Mint: 13 gallons
Great Grape: 12 gallons

JANUARY
Marvelous Mint: 9 gallons
Great Grape: 11 gallons

FEBRUARY
Marvelous Mint: 12 gallons
Great Grape: 12 gallons

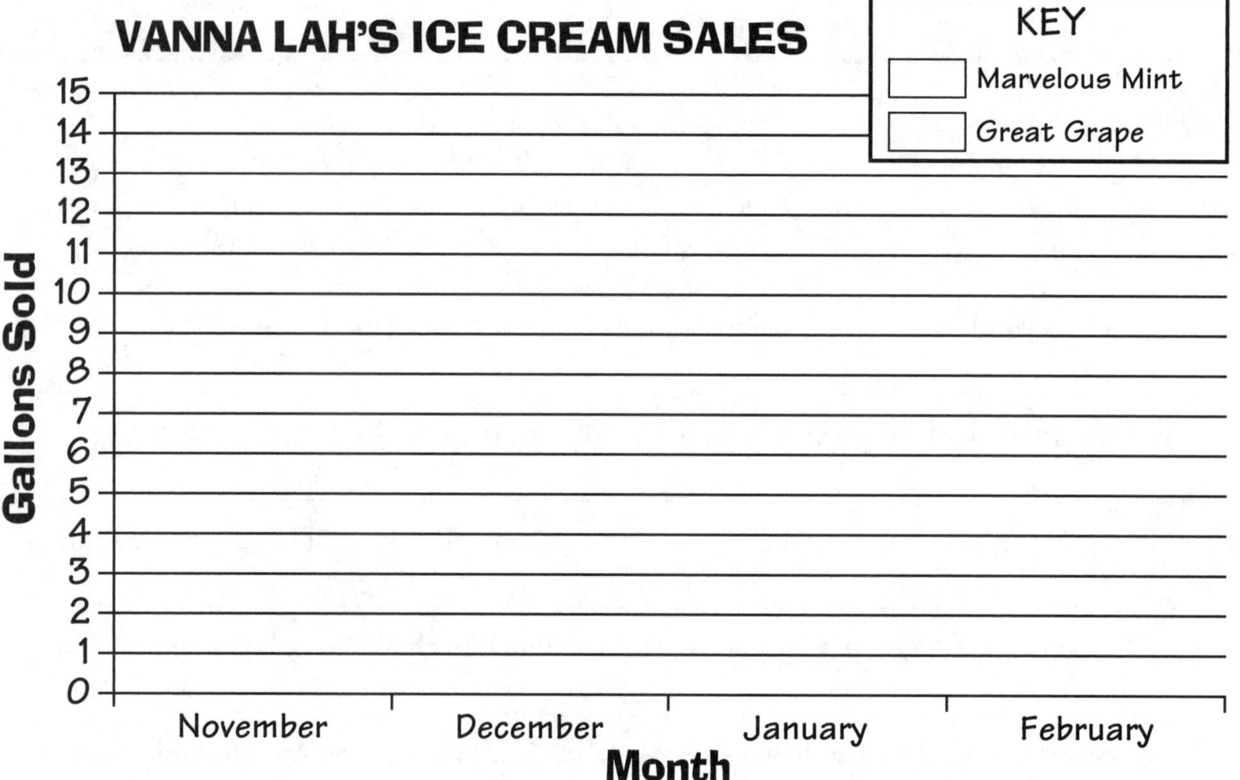

Decibel Tester

Name ______________________________ Date ______________

A bar graph is used to compare information.This bar graph shows the relative loudness of sound measured in decibels (dB). One decibel is the smallest difference between sound heard by the human ear. A 100-decibel sound is 10 times louder than a 10-decibel sound. A 100-decibel sound is painful!

JET PLANE TAKE OFF
SUBWAY
MOVIE THEATER*
THUNDER
LOUD ROCK MUSIC
NORMAL TRAFFIC
NOISY OFFICE
LOUD CONVERSATION
LIGHT TRAFFIC
NORMAL CONVERSATION
QUIET CONVERSATION
LOW WHISPER

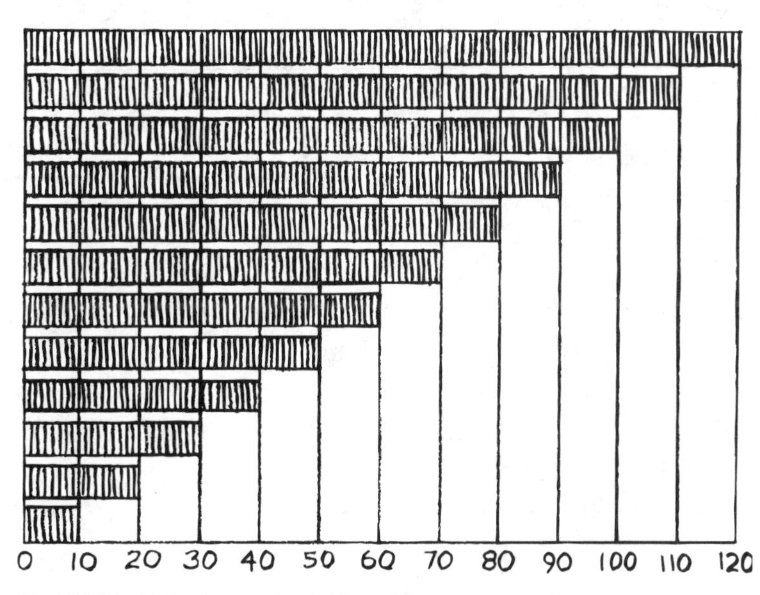

* SPEAKER VOLUMES IN A DIGITAL THEATER SOUND SYSTEM.

HOW MUCH LOUDER IN DECIBELS IS...

1. LOUD CONVERSATION THAN QUIET CONVERSATION? ____________
2. NORMAL TRAFFIC THAN LIGHT TRAFFIC? ____________
3. A MOVIE THEATER DIGITAL SOUND SYSTEM THAN LOUD ROCK MUSIC? ____________
4. A SUBWAY TRAIN THAN THUNDER? ____________
5. A NOISY OFFICE THAN QUIET CONVERSATION? ____________
6. WHAT IS THE LOUDEST SOUND SHOWN ON THE GRAPH? ____________
7. HOW MANY DECIBELS IS THE LOUDEST SOUND? ____________

Great Game Graph!

Name ______________________________ Date ______________

How much time do kids really spend playing video games? Read the circle graph to find out. Then, when you answer the questions, try to get the high score!

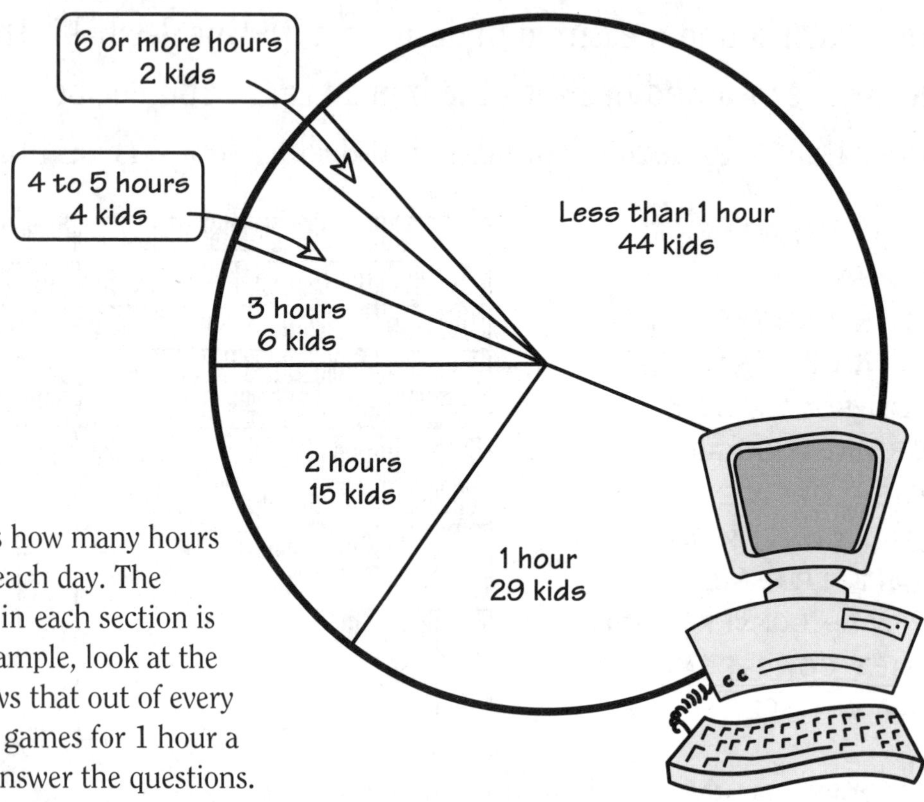

What to Do:

The circle graph shows how many hours kids play video games each day. The number of kids shown in each section is out of 100 kids. For example, look at the bottom section. It shows that out of every 100 kids, 29 play video games for 1 hour a day. Use the graph to answer the questions.

1. How many kids out of 100 play video games for 2 hours a day? ______________

2. How many hours a day do 6 out of 100 kids play video games? ______________

3. For how long does the largest group of kids play video games each day? ______________

4. For how long does the smallest group of kids play video games each day? ______________

5. Do more or less than $\frac{1}{2}$ of the kids play video games for less than 1 hour a day?

____

6. Think of the amount of time you play video games each day. What is the section of the graph where you would be? ______________________________

Brain Power

Do you think you and your friends spend too much time playing video games? Why or why not?

What's Hoppin'?

Name ______________________________ Date ______________

You Answer It!

1. Look at the graph below. Starting at square X, Judy hopped 4 squares up and 3 squares to the right. In which square did she land?

____

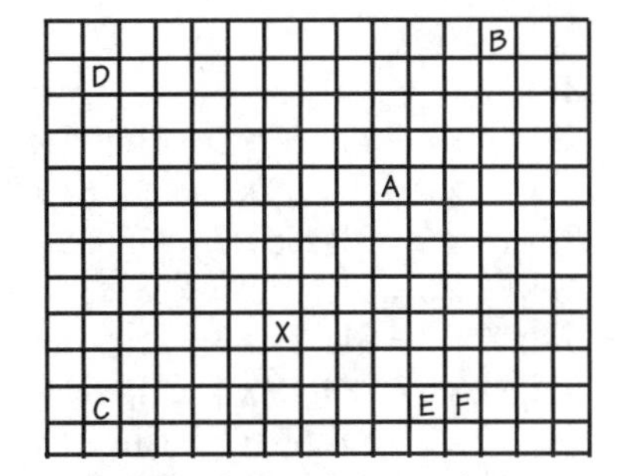

NOTE: Judy and Rudy can hop in vertical and horizontal directions only.

2. Rudy is in square X. Which are the 2 shortest paths he can take to get to square E? ______________________

____

3. Judy is in square A. Which are the 2 shortest paths she can take to get to square E? ______________________

____

4. Find the 2 shortest paths to get from square X to square D. ____________

____

5. Find 3 paths to get from square D to square E. Does each path contain the same total number of squares? ____

____

6. Starting at square X, Rudy hopped 6 squares up and 5 squares to the left. How many squares is he from square D?

____

Answer Key

Page 5

1. one; **2.** ten; **3.** eight; **4.** three
5. two; **6.** hundred; **7.** nine; **8.** thousand

Page 6

Winner - Karl Kat (16)
Second place - Sabrina Siamese (15)
Third place - Kelly Kitten (14)
Fourth place - Freddy Feline (13)

Page 7

1. 10; **2.** 9; **3.** 19; **4.** 25; **5.** 160
6. 7; **7.** 35; **8.** 15; **9.** 144; **10.** 400
What would you get if a pig learned karate?
Some pork chops

Page 8

1. thousands; **2.** 6; **3.** tens
4. 7; **5.** 0
6. billions; **7.** hundred millions
8. 727,912; **9.** 4,847,266
10. 7,446,732,011
How do skunks measure length?
In "scent"imeters

Page 9

1. BARK-RUFF, RUFF-BARK
2. You can make 2 phrases: BARK-GRR, GRR-BARK
3. You can make 6 phrases: BARK-GRR, GRR-BARK, BARK-RUFF, RUFF-BARK, GRR-RUFF, RUFF-GRR
4. You can make 6 phrases: MEOW-PURR, MEOW-SSS, PURR-SSS, PURR-MEOW, SSS-PURR, SSS-MEOW
5. You can make 6 phrases: MEOW-PURR-SSS, MEOW-SSS-PURR, PURR-MEOW-SSS, PURR-SSS-MEOW, SSS-PURR-MEOW, SSS-MEOW-PURR
Super Challenge: You can make 2 phrases: PURR-MEOW-SSS, PURR-SSS-MEOW

Page 10

1. 800; **2.** 5,000; **3.** 3,700; **4.** 1,000; **5.** 2,770
6. 8,000; **7.** 24,400; **8.** 11,000; **9.** 9,940; **10.** 73,000
What do cows give after an earthquake?
Milk shakes

Page 11

Answers will vary. You might encourage children to keep a list of their moves in order to defend their strategy.

Page 12

1. 100; **2.** 6,000
3. 1,013; **4.** 1,571
5. 4,247; **6.** 43,836
7. 15,033; **8.** 15,068
9. 1,000

Page 13

2	2	4	1			1	9	4	3				
		8		2	7	8			5	8	4	1	1
	1	5	7	8		4	1	1	8		2		3
5	0	2		9		0					2		3
	6			8		2	1	7	2		1	1	6
9			6		6			8		1			
0			5		8		2	3	3	5			
6	7	0	0		1			2		2	9	0	8

Page 14-15

Answers will vary depending on the numbers children select to start the game. Be sure that children check their numbers each time they land on a rooster. The number they have when landing on a rooster should be the same as the number they started with.

Page 16

1. 257; **2.** 428; **3.** 300; **4.** 743; **5.** 1,451
6. 2,869; **7.** 459; **8.** 48; **9.** 4,884; **10.** 7,926
What tables grow on farms?
"Vege"tables

Page 17

1. 4 x 71; **2.** 78 x 43 + 1; **3.** 143 x 8 – 7
4. 18 x 734; **or** 18 x 743; **or** 73 x 184; **or** 74 x 183
5. 418 ÷ 71; **or** 418 ÷ 73; **or** 471 ÷ 83; **or** 473 ÷ 81
6. 83 ÷ 17

Page 18

66 x 9 = 594; 94 x 8 = 752; 88 x 7 = 616; 91 x 8 = 728
98 x 8 = 784; 99 x 6 = 594; 68 x 8 = 544; 84 x 9 = 756
90 x 9 = 810; 89 x 6 = 534; 68 x 7 = 476; 38 x 2 = 76
95 x 5 = 475; 82 x 3 = 246; 97 x 6 = 582; 88 x 9 = 792
97 x 9 = 873; 92 x 7 = 644; 56 x 5 = 280; 81 x 4 = 324
68 x 9 = 612; 90 x 8 = 720; 79 x 4 = 316; 11 x 9 = 99
87 x 9 = 783; 99 x 9 = 891; 79 x 8 = 632; 74 x 9 = 666
79 x 9 = 711; 69 x 9 = 621; 82 x 9 = 738; 71 x 8 = 568
Taking It Further: a. 9; **b.** 7; **c.** 6

Page 19

1. 6; **2.** 8; **3.** 60; **4.** 63; **5.** 160
6. 252; **7.** 90; **8.** 56; **9.** 315; **10.** 96
How did the detective find the missing barber?
He "combed" the town.

Page 20

The only numbers not part of a multiplication number sentence are: 5 [row 1], 9 [row 5], 3 [row 6], and 11, 2, and 8 [row 8].

Page 21

1. 11,000; **2.** 24,000; **3.** 30,000; **4.** 56,000; **5.** 100,000
6. 144,000; **7.** 210,000; **8.** 256,000; **9.** 360,000; **10.** 375,000
Why did the spider join the baseball team?
To catch "flies"

Page 22
1. 2; **2.** 5; **3.** 8
4. 16; **5.** 3; **6.** 7
7. 13; **8.** 4; **9.** 9
Answer: IT'S "TULIPS"!

Page 23
1. 14 remainder 4; **2.** 2 remainder 8
3. 10 remainder 1; **4.** 5 remainder 5
5. 2 remainder 9; **6.** 3; **7.** 4 remainder 16
8. 9 remainder 1; **9.** 4; **10.** 8
What has 18 legs and catches flies?
A baseball team

Page 24
1. 4; **2.** 20 remainder 10
3. 11; **4.** 6 remainder 58
5. 9 remainder 7; **6.** 9 remainder 50
7. 5; **8.** 33 remainder 12
9. 30 remainder 40; **10.** 30 remainder 23
What has 3 feet but can't run?
A yardstick

Page 25
1. 56; **2.** 84; **3.** 95
4. 301; **5.** 250; **6.** 267

Page 26
1. 29; **2.** 10
3. 22; **4.** 60
5. Answers will vary.
Super Challenge: Answers will vary, but none of the products should be higher than 100.

Page 27

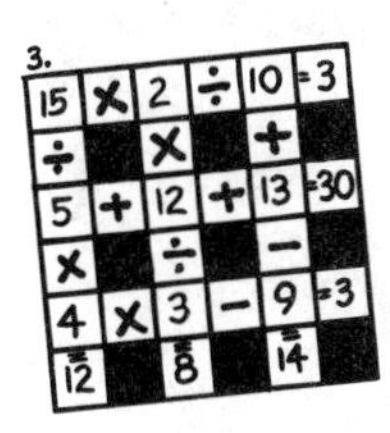

Page 28
1. 1/3; **2.** 3/8
3. 3/11; **4.** 14/20
5. 1/32; **6.** 20/67
7. 59/83; **8.** 22/121
9. 3/156; **10.** 99/312
What did one magnet say to the other magnet?
"You attract me."

Page 29

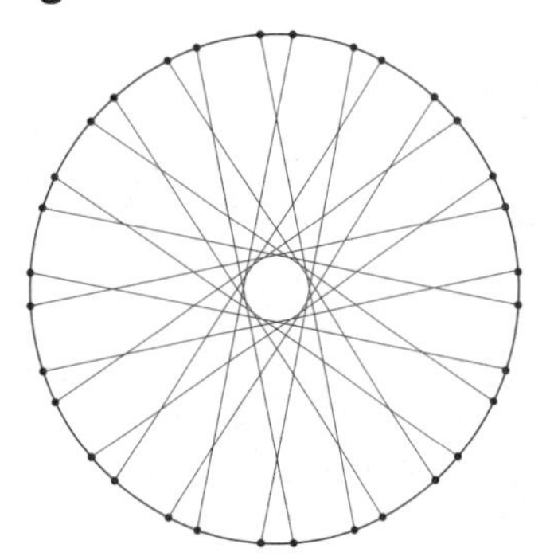

Page 30
1. 3/2; **2.** 5/3; **3.** 2/6; **4.** 8/5; **5.** 9/4
6. 14/8; **7.** 36/9; **8.** 10/3; **9.** 16/7; **10.** 54/11
Why did the artist need math?
He painted by numbers.

Page 31
1/2: 15/30; 9/18; 21/42; 11/22; 19/38; 7/14; 17/34; 50/100
1/3: 11/33; 3/9; 7/21; 14/42; 15/45; 14/42; 45/135; 9/27
1/4: 9/36; 7/28; 10/40; 25/100
Taking It Further:

$\frac{1}{7}$	$\frac{1}{7}$
$\frac{6}{14}$	$\frac{4}{14}$

$\frac{1}{2}$	$\frac{2}{14}$
$\frac{1}{14}$	$\frac{4}{14}$

$\frac{8}{14}$	$\frac{2}{14}$
$\frac{1}{7}$	$\frac{1}{7}$

Page 32
A. 0.6; **B.** 0.4; **C.** 0.42; **D.** 2.6; **E.** 2.6; **F.** 0.22
G. 0.25; **H.** 0.07; **I.** 3.2; **J.** 0.75; **K.** 4.6; **L.** 0.008
Answer: SO IT WON'T PEEL.

Page 33
Taking It Further:

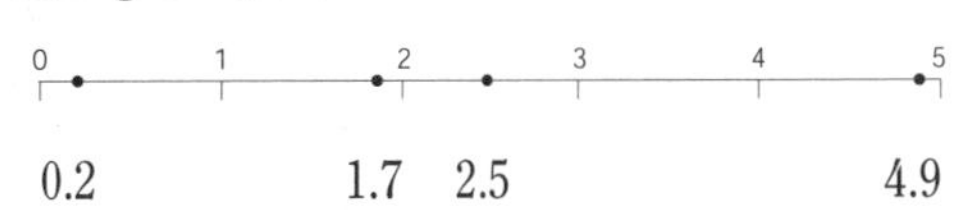

Page 34

Taking It Further: 0.70 = 0.7; 9.007 < 9.70, 0.30 = 0.3; 6.900 = 6.9; 0.90 > 0.09

Page 35

94.5 + 7.12 = 101.62; 28.5 + 71.12 = 99.62; 67.5 + 27.12 = 94.62
28.5 + 72.2 = 100.7; 76.5 + 21.12 = 97.62; 2.9 + 98.12 = 101.02
9.5 + 73.1 = 82.6; 24.5 + 7.12 = 31.62; 6.5 + 88.62 = 95.12
92.7 + 9.5 = 102.2; 2 + 79.12 = 81.12; 88.5 + 7.09 = 95.59
76.05 + 27.12 = 103.17
Taking It Further: 820.11

Page 36

5.2 – 4.1 = 1.1; 8.0 – 4.2 = 3.8; 9.24 – 2.9 = 6.34
6.0 – 5.7 = 0.3; 6.9 – 2.3 = 4.6; 96.1 – 65.8 = 30.3
9.9 – 2.47 = 7.43; 8.97 – 4.56 = 4.41; 0.78 – 0.4 = 0.38
6.0 – 1.1 = 4.9; 8.9 – 6.9 = 2.0; 3.8 – 1.7 = 2.1
8.9 – 3.1 = 5.8; 0.74 – 0.52 = 0.22; 2.9 – 1.6 = 1.3
7.8 – 3.1 = 4.7; 4.5 – 1.4 = 3.1; 5.3 – 2.2 = 3.1
8.6 – 2.0 = 6.6; 8.9 – 6.7 = 2.2; 4.47 – 1.04 = 3.43
9.62 – 1.1 = 8.52; 7.30 – 3.09 = 4.21; 6.6 – 2.0 = 4.6
5.6 – 3.5 = 2.1; 2.2 – 1.1 = 1.1; 9.8 – 5.1 = 4.7
2.7 – 1.7 = 1.0
Taking It Further: a. 2.391; **b.** 69.63; **c.** 76.03; **d.** 3.888

Page 37

186	Elway Electrician
7.2	Alan Architect
20	Carlton Carpenter
3	Paul Plumber
500	Robert Roofer
80	Gilbert Gardener
27	Penny Painter

Page 38

1. 8, 11, 14, 17, 20, 23, 26, 29
2. 27, 29, 31, 33, 35, 37, 39, 41
3. 2, 7, 12, 17, 22, 27, 32, 37, 42
4. 5, 9, 14, 23, 37, 60, 97, 157, 254
5. 39, 46, 53, 60, 67, 74, 81, 88, 95
6. 6, 7, 13, 20, 33, 55, 88, 143
7. 4, 15, 26, 37, 48, 59, 70, 81
8. 93, 116, 209, 325, 534, 859, 1393, 2252

Page 39

1. zero; **2.** times; **3.** factor
4. triple; **5.** double; **6.** product
7. multiply; **8.** addition; **9.** same size
Scrambled word: multiples

Page 40

1. 12; **2.** 8; **3.** 9
4. 31; **5.** 60; **6.** 36, 48

Page 41

1a. \$7.20; **b.** \$16.80
2a. \$6.80; **b.** \$10.20
3a. \$3.30; **b.** \$18.70
4a. \$13.00; **b.** \$52.00

Page 42

1. 3.5, 2.5; **2.** 4.25, 4
3. 0.6, 0.4; **4.** 4.8, 5.8

Page 43

1. 4 apples; **2.** 9 books; **3.** 25 caps; **4.** 20 oranges; **5.** 30 coats
6. 18 shoes; **7.** 56 bananas; **8.** 121 pencils; **9.** 54 video games
10. 42 chairs
How did the square become a triangle?
It cut a corner.

Page 44

1. 1.5 inches
2. 3.2 inches
3. The highest temperature was reached on Sunday. The lowest temperature was reached on Thursday. The difference between the two temperatures was 30 degrees.
4. The temperature dropped 15 degrees between Wednesday and Thursday. The temperature increased by 15 degrees between Saturday and Sunday.
5. 27 degrees
Super Challenge: 64 degrees

Page 45

1. Answers will vary.
Answers will vary.
6 cones
2. 5 cups
3. the cup
4. 5 1/2 drink boxes

Page 46

1. railway signal; **2.** ironing board
3. gas mask; **4.** cash register
5. ear muffs; **6.** ballpoint pen
7. hearing aid; **8.** windshield wiper

Page 47

1. Add four crackers (units).

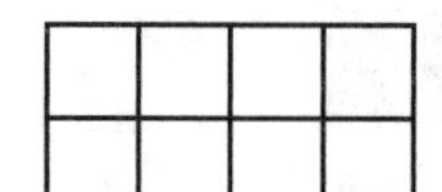

2. The area will be 9.

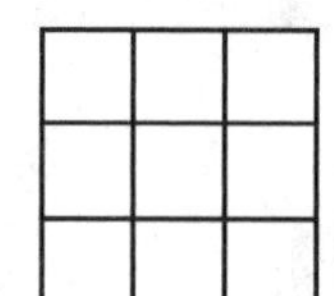

3.

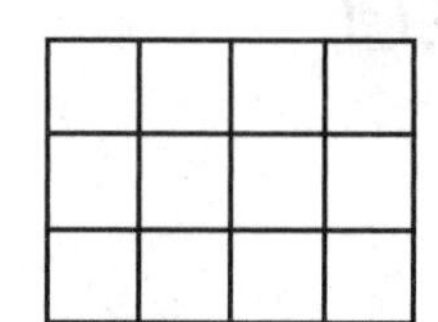

4. Possible answers include: a square 4 units by 4 units with a perimeter of 16; a rectangle 2 units by 8 units with a perimeter of 20; a rectangle 1 unit by 16 units with a perimeter of 34.
5. Answers will vary, but any correct answer will use 24 units.
Brain Power: One possible shape is a square made of 16 units.

Page 48

1. 12; **2.** 24; **3.** 48; **4.** 60; **5.** 84
6. 108; **7.** 120; **8.** 72; **9.** 144; **10.** 180
Why did the sick book visit the library?
To get "checked out"

Page 49

1. 60; **2.** 120; **3.** 240; **4.** 300; **5.** 420
6. 600; **7.** 660; **8.** 900; **9.** 1,080; **10.** 1,200
Where does a king stay when he goes to the beach?
A sand castle

Page 50

1. 9 a.m.; **2.** 3 p.m.; **3.** 9:30 p.m.; **4.** 7 a.m.; **5.** 11:30 a.m.
6. 8:20 a.m.; **7.** 2:17 p.m.; **8.** 10 p.m.; **9.** 8 p.m.; **10.** 1:11 p.m.
What part of a cowboy is the saddest?
His blue jeans

Page 51

1. 16; **2.** 20; **3.** 19; **4.** 40 inches; **5.** 12 inches
6. 9 inches; **7.** 55 inches; **8.** 30; **9.** 38 inches; **10.** 74 feet
Why do people with colds get plenty of exercise?
Their noses run.

Page 52

1. 5 units; **2.** column 7; **3.** column 3; **4.** 20 units
5. columns 1 and 5; **6.** 25 units; **7.** 15 units
8. 30 units; **9.** 40 units; **10.** column 4
How can you dive without getting wet?
Go skydiving.

Page 53

1. The canary's heart rate is the fastest. The gray whale's heart rate is the slowest.
2. Answers may vary. In general, the smaller the animal, the faster the heart rate.
3. A horse's heart beats 25–40 times per minute. It would fit between the elephant and the tiger. That answer would be logical because a horse is larger than a tiger but smaller than an elephant.
4. Answers will vary.

Page 54

4. Answers will vary.
5. Answers will vary.

Brain Power: One way to answer this question is first to estimate or measure the number of cups of cereal in the full box, then to multiply the number of each letter you counted in one cup by the number of cups in the box.

Page 55

1. Dodge Stadium and County Stadium; $4.25
2. $1.25
3. $3.50; It is the price that appears most often on the graph.
4. Answers will vary. Possible answers include: the amount of profit the vendors need to make; the price of hot dogs in the area; how much it costs to buy and make the hot dogs.
5. Answers will vary. Students might suggest they would want to know the cost of tickets, souvenirs, other food, parking, and transportation to the stadium. To find these costs before the game, they could call the stadium or ask someone who has been there before.

Page 56

Students' graphs should represent the sale of the two flavors for the four months given.

Page 57

1. 30 dB; **2.** 30 dB; **3.** 20 dB; **4.** 20 dB
5. 40 dB; **6.** Jet Plane Take Off; **7.** 120 dB

Page 58

1. 15 kids; **2.** 3 hours a day
3. Less than 1 hour a day; **4.** 6 or more hours a day
5. Less than 1/2; **6.** Answers will vary.
Brain Power: Answers will vary.

Page 59

1. Judy landed in square A.
2. Down 2 squares and 4 squares to the right. Four squares to the right and down 2 squares.
3. One square to the right and 6 squares down. Six squares down and 1 square to the right.
4. Seven squares up and 5 squares to the left. Five squares to the left and 7 squares up.
5. Answers will vary. Each path does not have to have the same number of squares.
6. He is 1 square down from square D.

Instant Skills Index

SECTION AND SKILL .... ***PAGES***